Losing The Sky

Losing The Sky

The threat from satellite mega-constellations,
why it matters,
and what we can do about it

Andy Lawrence

with assistance from

Jonathan McDowell
Robert J. Antonucci

Foreword by Brian May

Andy Lawrence
Institute for Astronomy, University of Edinburgh
Royal Observatory, Edinburgh EH9 3HJ

Published by: Photon Productions, Edinburgh
February, 2021

This work is available in various formats as follows:
ISBN: 978-1-8383997-0-2 (EPUB)
ISBN: 978-1-8383997-1-9 (Kindle)
ISBN: 978-1-8383997-2-6 (paperback)

Cover image: Rafael Schmall, used with permission.
Credits for other images are given in the text.
Where no credit is given, the picture is by the author.

Half of the net income from this book will be donated by the author to the Royal Observatory (Edinburgh) Trust:
http://www.roetrust.org.uk

Praise for *Losing The Sky*

"The rapidly increasing problem of numerous satellites streaking across the night sky threatens to disrupt astronomical research, which is in a golden age of discovery. Professor Lawrence's book ought to be required reading for every astronomer and astronomy enthusiast who values the starry sky and what might happen to it in future times."

Dave Eicher, Editor in chief, Astronomy magazine

"They say that there's no such thing as a free lunch. The same applies to "cheap internet access", especially when it comes from space. As commercial developers flood low Earth orbit with satellites to exploit the demand for more movie streaming, videoconferencing, low-latency gaming, and high-speed trading, what's the real cost? Are we willing to sacrifice the night sky as thousands of artificial objects constantly streak overhead, changing the natural heavens for astronomers and the wider public alike without escape? Andy Lawrence's book, "Losing the Sky", cogently summarises the issues and the concern that the world may be unwittingly sacrificing a public commons, our fundamental common heritage of a portal on the Universe, on the altar of Mammon. It's time to stop, think, and decide whether we're really willing to pay that price."

Mark McCaughrean, Senior Science Adviser,
European Space Agency

"An important book about an important subject. The night sky is part of our natural heritage, and here Professor Andy Lawrence explains how obliteration of our view of the heavens by the next generation of communications satellites is akin to destruction of the Earth's natural habitats. He's persuasive on why we should all - not just astronomers - care about this loss, and what we can do about it."

Pippa Goldschmidt, author of "The Falling Sky" and "The Need for Better Regulation of Outer Space"

"Sixty years ago, Rachel Carson published Silent Spring, alerting the nation and the world to the widespread dangers posed by toxic chemicals in the environment to nature and humankind, So influential was that landmark work, that President Kennedy took the first governmental measures to regulate dangerous artificial chemicals. Similarly, Professor Lawrence has alerted readers to the existential threat posed by large-scale satellite projects to humankind's oldest and most treasured heritage — the heavens. It is a conflict between what Hunter S. Thompson would have called the "Greedheads" — individuals of enormous wealth who wish to profit from space—and the universal human right of people for an unobstructed view of the stars. Simply put, this remarkable book should be in all libraries, public or private, that aspire to be complete on the subjects of science and technology. Working together, people around the planet, inspired by this eloquent and passionate book, can put an end to the madness and folly of "artificial constellations" "

John A. Murray, author or editor of 45 nature books, including "Wild Africa", "A Republic of Rivers", and "The Islands and The Sea"

Table of Contents

FOREWORD

The end of Astronomy as we know it ?

PROFESSOR LAWRENCE, IN THIS TIMELY BOOK, tackles an issue which is about to become highly contentious around the world. The rate at which satellites are being placed in orbit around the Earth is currently exploding, to the point where, as thousands of them streak brightly across the night sky, they will make ground-based astronomy and astro photography almost impossible. The case for and against this 'progress' is thoroughly discussed here, and, as in many modern issues, Lawrence reveals that there is an array of consequences which have not been fully assessed before the button was already pressed on this new horde of man-made satellites. This book will hopefully spark enough discussion to put the brakes on this destruction of our dark skies.

Brian May
Jan 26th 2021

PREFACE

THIS BOOK HAS COME ABOUT through a mixture of horror, excitement, and pressure. Like a lot of astronomers, I was shocked when I first saw images taken by my professional and amateur colleagues that were ruined by streaks caused by the new SpaceX Starlink satelllites. We were used to the occasional satellite, but this was something else. How did we not see this coming? I began to think about how to help my community do something about it. Join one of the working groups being set up? Write a New Scientist article? Give some public talks?

Then, just after the SATCON-1 report came out, I saw a Facebook post by the Californian astronomer Robert (Ski) Antonucci. We both work on quasars and black holes, and have spent many years swapping strong views on accretion discs and obscuring regions. Ski was incensed by the satellite mega-constellation business. He suggested that what was needed was some kind of polemical book, along the lines of Silent Spring or The Rights of Man, and was saying "who will do this?". At that moment, I was halfway through drafting a semi-popular book on Mountain Astronomy and finding it exciting but difficult. I found myself thinking "well, I am not exactly Rachel Carson, let alone Thomas Paine, but I think I could do this". So I messaged Ski and said yes, and since then we have been communicating several times a week. The writing and research is all mine, but Ski has been constantly pushing, suggesting, and arguing. He is good at this.

Meanwhile, Ski said he had a secret weapon - the best possible technical consultant. This turned out to be Jonathan McDowell, aka "Planet 4589". This was good news, both because Jonathan is possibly the acknowledged expert on whats up there and why, and because we are old friends, having known each other since the 1980s. And like myself and Ski, he too writes papers on quasars. I send Jonathan drafts to check their accuracy, call him up for the occasional explanation, and also get data from him for making diagrams. Jonathan is an interesting character in this story, because he is a keen and knowledgeable space nerd, speaks to people in SpaceX, and so on. He is the polar opposite of a Luddite, but still deeply concerned about satellite constellations and "New Space".

As I started to plan the book and daydream about what I needed to write, what emerged was maybe not quite what Ski envisaged. I was struggling to understand what this meant to me, and why anybody else should care. I was worried that I would look like a Luddite, resisting progress. I was worried that my community would look like just another lobby group, pleading that our area is special and must be protected. I was worried that people would think this was just a minor problem, compared to climate change and the loss of biodiversity - and indeed those are the big problems facing the world. Then it struck me. The satellite problem, and the longer standing problem of light pollution, are *part* of our ongoing environmental catastrophe. Losing the sky is just the last damned straw. This is why everybody should care.

I didn't want to write a detailed technical work - other people are already doing much more important and thorough things than me on that front. But I also didn't want to write a kind of political manifesto or campaign script. A manifesto is probably needed, but this isn't it. So what did I have to contribute? I felt that basically I needed to write from the heart, combining my personal experience with my scientific knowledge, and explain to ordinary people what was going on, why it mattered, and roughly what our options are. So the book is written in a very concrete, personal, and informal style, while trying to get across big social, cultural, and technical issues. This is a strange combination, so I hope it works for you. For those who want more detail, the Appendix has lots of links to the web pages, videos, books, and technical articles that I have consulted.

Human activity, enterprise, technology, and commercial development are all wonderful - but unregulated and unrestrained they can cause serious problems. This is not necessarily about greed or bad behaviour. (Athough sometimes it is!) It's about the tragedy of the commons. People use up things that seem to be free, or leave mess for "society" to clean up. But such behaviour is a cost for all of us. Dumping greenhouse gases in the atmosphere is the big example of course. But the sky is another. Hey, space is free, lets just use it!! Wrong. Within minutes, your new satellite is polluting the sky in parts of the world whose permission you never asked.

The sky belongs to all of us. Lets not lose it.

Andy Lawrence
Royal Observatory Edinburgh
February 1st 2021

I. THINGS FALL APART

September 17th, 2020

THIS IS THE STORY of how we are in danger of losing the sky forever. Let me explain what it means to me, and how tragic this seems.

THE YEAR IS 1963. I am nine years old. Kent is in the grip of the worst winter in many years. I have an incomplete but vivid memory. Our car has broken down, somewhere in the middle of the sea of cabbages that separates Margate, Ramsgate, and Broadstairs. I am with my Dad and big sister Sue. Mum is at home with little sister Deb. They will be worried, but there is no way to let them know - no mobile phones, no internet, in 1963. I can remember very little else. Why were we there? Where had we come from? How did the story end? I don't know. I just have two pictures in my head.

First, the snow. Piled up beside the car. Cold, white, and beautiful. As the late afternoon sun sank low, the snow glistened in the sideways light.

The second picture burned into memory is this: the stars. As the sun sank lower still, the world underwent a magical transformation. The sky darkened in coloured stripes, and one by one, points of light mysteriously appeared. An hour later, the sky was inky black, spangled with thousands of stars. Like the snow, the sky sparkles in my memory.

FIVE YEARS LATER. 1968. My favourite books are all about space and the Universe. I read about giant telescopes in California, reaching deep into space, utterly changing our understanding of the Universe we live in - galaxies, quasars, the expanding Universe and the Big Bang. At about this same time, I team up with school chum Paul Sutherland and others, forming the Thanet Astronomical Society for Youth. I think I am a disappointment to them. They are proper keen amateurs, counting meteors and observing clusters and all that sort of thing. I am frankly an armchair astronomer. I am on an intellectual journey, but also a romantic one - my head is full of galaxies and rockets and giant telescopes in far away places.

ROLL FORWARD ANOTHER ELEVEN YEARS. 1979. I am living the dream. I have studied Astrophysics in Edinburgh, and have then moved to Leicester to do a PhD in X-ray astronomy. This is hugely exciting, but still a kind of armchair astronomy. My data comes from the Sky Survey Instrument on Ariel V, a spacecraft orbiting 500 km above us. All I see is the stream of bits arriving in the Leicester data room.

But now I am on a mountain in Tenerife, and re-connecting with the physicality of my work. Like many X-ray astronomers of the time, I need to observe the sky in visible light, and spot the object that those X-rays are coming from. Actually, I am going one step further, and observing my favourite galaxy (NGC 4151) in the infrared, using the 60inch flux collector on the Izana ridge on Tenerife. It all feels much more real. I had to drive up a mountain, sleep in an uncomfortable bed during the day, stare through an eyepiece, and push buttons to guide the telescope, while leaning awkwardly on a step ladder.

Suddenly there is a power cut. The telescope drive stops, and the control system has lost all knowledge of where it was pointing. Step one - I trudge down the road in the dark to a small building where I restart the generator. Step two - I get out Norton's Star Atlas, and then stare at the sky to find Alpha Boo. Step three - I manually push round the telescope and get Alpha Boo on the eyepiece cross hairs. Step four - I key in the co-ordinates and restart the drive. This whole procedure is known as a SNAFU. Don't make me spell that out... Hard work, but gosh I feel alive, and somehow it drives home that science is not an abstract game. My favourite galaxy is not something on TV - it is really there.

DECADES LATER, I AM BACK IN TENERIFE. The year is 2016. I am involved in a film about the adventures of my illustrious predecessor Charles Piazzi Smyth. In 1856, he started the idea of mountaintop astronomy, living on the mountain for two months with his geologist wife Jessica and a bunch of sailors, proving that the skies were clearer, darker and sharper, when you get above a third of the Earth's atmosphere. Now we have gathered some Scottish and Spanish students, and are on Tenerife trying to duplicate Piazzi Smyth's classic experiments - on the solar spectrum, on double stars, and on the Zodiacal Light. Its being filmed by

Guillermo Carnero-Rosell, to make a movie called "A Residence Above The Stars".

Today we have climbed Mount Guajara, a three hour trek from the Parador in the Caldera, and found the remains of Piazzi Smyth's stone house. A few km away, on Izana, there is a veritable village of telescopes and infrastructure; here on Guajara there is nothing but rocks and sky. Teide looms across the Caldera. We are here with Brian May, the well known astronomer turned musician turned astronomer again. Brian started his PhD on Tenerife, studying the Zodiacal Light, and knows all about Piazzi Smyth. He has wanted to ascend Guajara for years.

As twilight falls, Brian shows us the Zodiacal Light. He knows his stuff. The Zodiacal Light is a very faint glow, caused by dust particles in the plane of the solar system scattering sunlight. It is very hard to see, and I have never managed before. But in the beautiful skies of Guajara, it all becomes clear. The sky is almost completely dark, and Brian knows what direction the nearest town is, so we don't confuse the two faint glows. For a few more hours, we wander around the sky, looking at double stars with our binoculars, finding the Andromeda Nebula, talking, laughing. Then we settle down in our sleeping bags and sleep under the naked sky. Its achingly beautiful - the stars are sharp, the Milky Way wheels above me, and I see a shooting star every few minutes.

SO NOW ITS 2020. What happens as I look back on these memories? Well, where the car broke down in 1963 is now a shopping mall, and cars and lights are everywhere. Would I still see stars and a magical transformation? Not a chance.

What about California, where our understanding of the Universe was transformed? Take a look at Figure 1.1 - a recent picture taken from Mount Wilson, where so many historic observations were made. You can see a double environmental disaster. Forest fires are destroying the land, a likely result of climate change due to the activity of humanity. But also, the view from Mount Wilson is dominated by the lights of Los Angeles. A beautiful sight in its own way, but the stars stand no chance in that city glare.

Figure 1.1: Mount Wilson, September 2020. Picture from the HPWren camera system at Mt Wilson Observatory.

There are still dark places, although they are getting rarer. Professional astronomers can put their telescopes on remote mountaintops, where skies are dark and the images are sharp. But a large fraction of humanity has already lost their connection to the sky. You are lucky if you live somewhere dark, or can visit a National Park. There is an international effort to try to preserve areas as "Dark Sky Parks" but they don't have any kind of formal legal protection.

In 2016, Guajara at least was a wonderfully dark place. I haven't been back, but I am sure it still is. However, there is a new problem, from which there is no escape anywhere on the surface of the Earth. One of the stars we explored with our binoculars that night on Guajara was Albireo, otherwise known as Beta Cygni. Every year, the Royal Observatory at Greenwich runs a competition for "Astronomy Photographer of the Year". One of the entries this year featured a picture of Albireo. You can see it in Figure 1.2.

What are those streaks? What is going on? Albireo

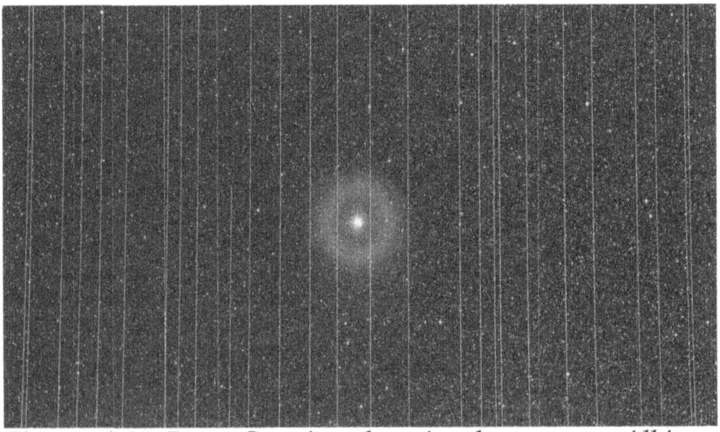

Figure 1.2: Beta Cygni, otherwise known as Albireo, photographed by Rafael Schmall, with Starlink satellites streaking across the image.

appears to be imprisoned behind bars! What we are seeing is streaks from a constellation of satellites launched by the commercial company SpaceX. So far, along with satellites by another company called OneWeb, there just a few hundred of these; but soon there will be thousands. SpaceX and OneWeb have started to talk to astronomers about how to "mitigate" the damage caused - but its not obvious their competitors will do the same. Later we will see just how bad this problem could become. But why is this happening?

II. ONLY CONNECT

October 4th 2020

WHY? WHY SHOULD ANYBODY SURROUND THE EARTH with thousands of satellites? Shortly, we shall ask why we should care, and what nightmare future could unfold. But first let us try to understand why we might want to do it at all. Its about the human desire to stay in contact, every hour, every second of the day. It's a story of empowerment, freedom, and democracy; of entertainment, drama, and gaming; of greed and the power of capital. It's a story where the appeal of outer space has waxed and waned and is now waxing again.

FOR THOUSANDS OF YEARS, there have been two ways to send a message to a distant place. Method number one is to open up a dedicated connection and send a signal across open space - wave the flags, flash the lantern. The message gets there very fast, but you have to know the other person is watching before you start. Method number two is to write a letter and put the address on the outside. The letter is passed from servant to rider to ship to coach, in gradual, unplanned steps. The intended recipient doesn't even know the letter is coming until one day the postie brings it up the path. Method number two is slow but easy.

Figure 2.1: Artist's impression of the thousands of satellites that are coming our way. Image credit: NOIRLab/ NSF/ AURA/ P.Marenfeld. (From the cover of the SATCON1 report - see Appendix)

In the nineteenth century human connection was transformed by two things: electricity and the train. The post got faster, but the point-to-point flashing of signals got even faster still - a telegraph signal could be transmitted at electric speed to anywhere you could string a cable. Next, we worked out how to turn our voices into coded electrical pulses, and then how to transmit signals by radio waves, at the speed of light. Soon the world was connected by an intricate spiderweb of wires, cables and radio masts. This communications infrastructure was impressive and powerful but hellishly complex - physically, legally, and financially. Who owned what, and who did you pay to connect?

Then in 1957 a tiny tin can was shot into space and the Space Age began. All Sputnik did was beep at us, but that one tin can could send its beep to anywhere on Earth, as once every eighty four minutes it passed overhead. I was only three years old, but as I grew up over the next decade I was swept along with the excitement of the new world - humanity was connecting up, and our future was in the stars. By 1962 Telstar could transmit images and telephone calls. The Telstar trick was to get up to a higher orbit, where its beamed data could reach more of planet Earth. Each ground station could beam data up, or read it back down, for twenty minutes at a time until the satellite sank over its horizon. Then in 1967, the summer of love, came the first live global TV show. The Beatles sang to the world - All You Need Is Love. Two years later Neil Armstrong stood on the Moon and we watched him on TV in the comfort of our homes. The future had arrived. Those grainy TV pictures arrived with a delay of just over a second - thats how long it takes radio signals to travel from the moon - but hey, nobody is going to worry about such a short delay, right? Watch this space.

TODAY, TELSTAR-1 IS STILL UP THERE, quietly orbiting. There have been more Telstars (Telstar-18V was launched by SpaceX in 2018) as well as numerous other communications satellites. Spacecraft do many useful things for us - monitor the weather, observe the Earth, pin down our location, and spy on our enemies. But they did not become the dominant force in global connectedness. Why not? Answer-1: The Internet. Answer-2: Optical Fibres.

The Internet works by the pass-the-letter-on method, moving packets of data from one computer to the next until they get to their intended addresses. But the network of roadways the Internet used was largely the same as the intricate complex of cables, wires, and phone masts that had already been developing for decades. This infrastructure has gradually been upgraded and replaced, with the Internet now in mind, by optical fibres - both the last-mile feed to our homes and offices, and the main "trunk roads" under the ocean carrying data traffic from one continent to another.

Optical fibres have two enormous advantages - bandwidth and latency - or in other words, volume and speed. The coded pulses travelling down one single fibre can in principle carry Terabits of data every second - enough for millions of pictures. Communications satellites fall thousands of times short of this volume of data transmission. As we all go Internet crazy - watching movies, making video calls, and playing online games - the ability to shift data in bulk has come to really matter.

So what about speed? Originally this was a big advantage for the satellites, as the radio waves they used beamed down data at the speed of light. Now the cables on the ground have caught up, as they send signals with light bouncing along the optical fibres. The crucial thing is the time delay or lag. The latest undersea cables can send a signal from London to New York with a delay of 59 milli-seconds. Now, communications satellites tend to be in so-called geo-stationary orbits, high above the Earth, because they can always be seen, and in turn can see most of the Earth. However the time lag from a geo-stationary satellite to the Earth is an eighth of a second.

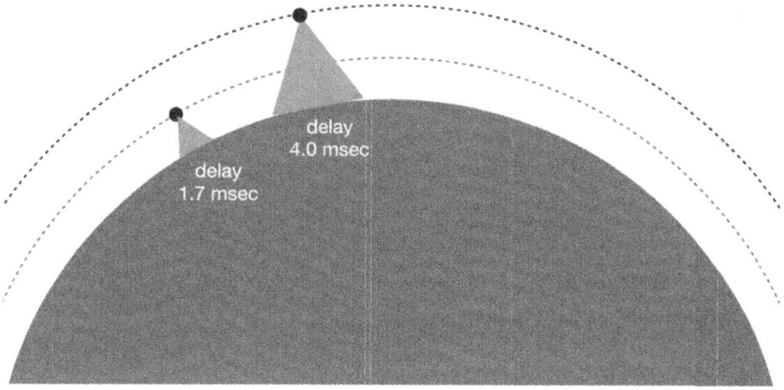

Figure 2.2: A low orbit satellite can only beam down to a small patch of the Earth, but the travel time down to Earth (the latency) is very short. Here we illustrate orbits at heights of 500km and 1200km

Who could be so impatient as to worry about a lag of a eighth of a second? Surely anybody can wait that long for a web page to load? Even if you are passively watching a streaming movie, you would surely never know the difference? True, but it starts to matter if you are not passive, but want to react. Here are two contrasting examples. Gamers are play-fighting each other across continents. You need to react fast when you see that rival avatar raise their sword; if there is a quarter second there and back delay, you don't get a smooth realistic experience. Exchange dealers are trading all the time as they see the prices fluctuate, trying to cream money off by buying and selling at exactly the right time. This is now done largely by automated algorithms; if you know the stock prices milli-seconds before your rivals, you stand to make millions of dollars. Several of the most recent transatlantic cables, with the fastest speeds, have been paid for by private consortia of investors, going straight from data centres in London to data centres in New York.

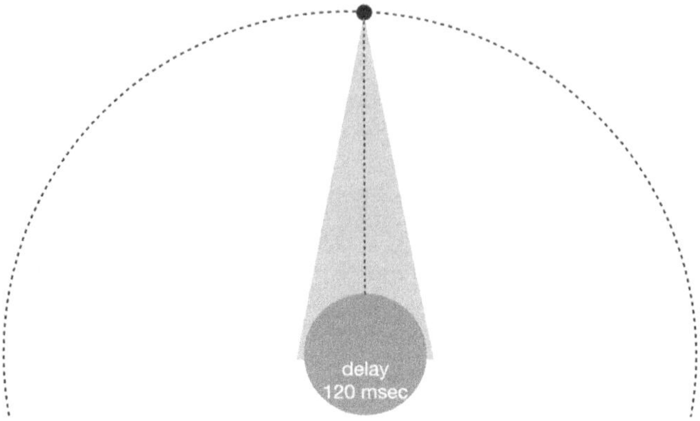

Figure 2.3 A satellite in a high orbit, such as a geostationary orbit, can see half the Earth, but has a long delay in beaming down.

ONE MORE KEY ISSUE about the current communications infrastructure, and then we are ready to understand what the new satellite constellations are all about.

The network of cables, fibres and masts has spread like a tangled weed across the modern world, but it doesn't go everywhere. Large parts of the world are poorly served. This matters a lot to some people with deep pockets - oil companies with ocean rigs, or armies operating in remote locations. Potentially it is also of interest to billions of less affluent customers. We tend to assume that the Internet has transformed all our daily lives; but in fact 60% of the world has no Internet connection, and even more of the world has a connection considered unsatisfactory by dedicated gamers, let alone financial wizards.

So now at last we can see the opportunity for the new constellations of satellites. It's not about bulk volume of data - the satellites won't beat the fat ground infrastructure in crowded cities. Its about global access - reaching everywhere - and about long-distance latency - reducing the lag for those excited gamers and the frantic traders. Traditionally, to connect to satellite Internet data, you need some pretty sophisticated and expensive equipment. This is fine for those oil companies and militaries, but no good for ordinary people across the world. For global access to work, ordinary people have to be able to connect to the satellites, at a reasonable price. SpaceX claim to have cracked this problem, with plug and play kit that is akin to a TV satellite dish. Their initial product is really not that cheap (see the Appendix), but maybe it will come down in price. Of course you do have to make sure that your satellite is always in the sky anytime you look up. Low Earth Orbit satellites, the cheapest to launch, are not so good at this. Their height, a few hundred km up, is much less than the radius of the Earth, so as you watch, they pretty soon disappear over the horizon. You can look at this from the point of view of the satellite. A Low Earth Orbit satellite, as it looks down, can see only a small patch of the surface of the Earth. A satellite in a high orbit on the other hand can see a large fraction of the Earth (see Figures 2.2 and 2.3).

So if you want to be connected all the time from anywhere, you can do it one of two ways. Either you can have a handful of satellites in relatively high orbits, or you can have thousands of satellites in low orbits. The high orbit version has already been done. The telecommunications company Viasat already offers high-bandwidth Internet to rural locations in the USA, from a single geostationary satellite. The company O3B (which stands for "the other 3 billion") offer global internet access, using a modest number (20) of satellites in Medium Earth Orbit, about 8000km up.

So why would you want to go the low orbit route?

Doesn't it seem crazy to build thousands of satellites rather than a few? Well, not if you are obsessed with that long distance delay. Imagine getting a message from Tokyo to New York. With those high orbit satellites you have to bounce the message from ground station to satellite and back to another ground station, maybe two or three times, with every leg costing a tenth of a second or more. If on the other hand you have thousands of low orbit satellites, the trick is to pass the signal from one satellite to its neighbour, and then beam it down on request, a very short final hop. The height of the satellite - say 500km - is trivial compared to the distance around the globe from Tokyo to New York - about 11,000km - and much less still than the distance up to a geostationary satellite - about 36,000km - so you can get much shorter lags with a connected network of low orbit spacecraft.

But is there any advantage for low orbit satellites over those zippy fibre optic cables? Yes. Advantage number one is that the speed at which light travels through glass, fast though it is, is two thirds of the speed of light in a vacuum. So passing the signals across empty space gives you a big win. Of course, you can throw that advantage away if you make the whole signal journey longer; so you need to keep those satellites low, which in turn means you need many many satellites in a connected chain. Advantage number two for the satellites is that you don't need to lay underground fibre all over the world - a very expensive business. You just ask any customer who wants the service to buy a modest extra piece of kit

.

WHAT WE HAVE ARRIVED AT - a constellation of thousands of low orbit satellites - is a beautiful and visionary scheme. But it is hard to imagine anything better designed to foul up the night sky. The satellites can be as bright as the stars you can see with your naked eye, and once the scheme is complete, there will be many more satellites than stars. As well polluting the sky for ordinary folk, they will streak across the cameras of professional astronomers, saturating sensitive detectors. The potential for collisions and the piling up of space debris also increases horribly.

Surely there are already lots of satellites up there? How can this be so much worse? Its true that the number of satellites has climbed inexorably over decades, but there has been a dramatic rise in the last two years, especially since mid-2019, when SpaceX started launching the Starlink satellites.

In 1990 there was something like two hundred satellites; by 2015, around a thousand; by late this year, 2020, there are about three thousand - of which 817 of are Starlink satellites. Based on applications by a variety of commercial companies, by 2029 we expect something like a hundred thousand.

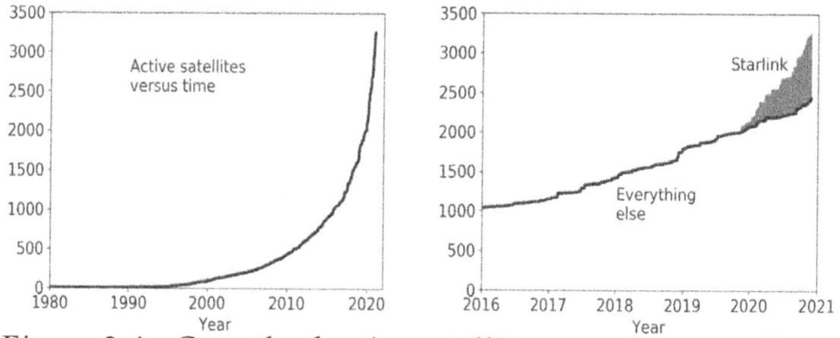

Figure 2.4 : Growth of active satellites over the years. Data provided by Jonathan McDowell.

The scariest thing is that most of these new and expected satellites will be in Low Earth Orbit, and so will be bright in the sky. Satellites shine by light reflected from the Sun, just like the Moon does. How much light they reflect depends on how big they are, but how bright they look to us here on Earth also depends on how high up they are. Those distant geostationary satellites can only be seen with a decent sized telescope; but the Starlink satellites, at 600km, are as bright as the stars we can see with our eyes. Some proposed satellite constellations would be higher, around 1200km. If they are the same size as the Starlink satellites, they would be somewhat fainter, perhaps just out of naked eye reach, while still a problem for amateur stargazers and professional astronomers. But note that "if". Higher but bigger satellites would still be a serious problem.

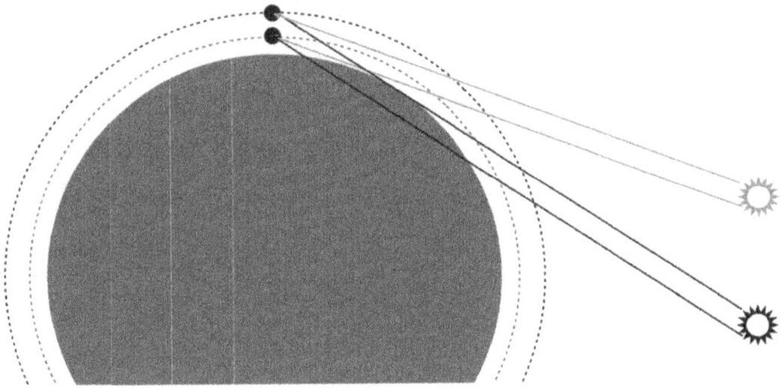

Figure 2.5 : Satellites can see the Sun when we can't, and satellites in higher orbits can see the Sun for longer as it sinks below the horizon.

Are these satellites visible all night? This also depends on their height above the Earth. They are up there all the time, but only look bright when they can see the Sun, and reflect its light. Because they are high above the Earth, they can still see the Sun when, from our point of view, the Sun has sunk below the horizon; but as the Sun sinks further still during the night, even the satellite can no longer see it, and the satellite goes dark. A satellite that is higher above the Earth can see the Sun for longer.

First, suppose that you are observing the sky from low

latitudes (say 20 or 30 degrees North or South), where indeed a lot of professional observatories are. Those low orbit satellites might be visible for only an hour or two after twilight. So what's the problem, the Starlink fan might say - just wait. However, early evening is exactly the time when most ordinary people look at the sky, and many key types of astronomical observation need to take place. (See Chapter 5). Slightly higher satellites, at say 1200km, will be fainter, but visible much longer.

But now consider someone just like myself, living at 56 degrees North. There are many of us, not lucky enough to live in Hawaii... The Sun never gets very high in the day... but also never gets very low during the night. On Sept 23rd, those 600km satellites are visible for five and a half hours out of the ten and a half hour night, and the 1200km satellites are visible all night; on July 15th the night is only five hours long, and even the 600km satellites are visible all night.

SURELY SOMETHING CAN BE DONE to make the satellites less damaging to the sky? Well yes; and the folk at SpaceX have already engaged in conversations with professional astronomers, after the first burst of fuss, trying out black paint and shields, and cunning deployment plans. These things will make some difference, though probably not enough. But what happens when greedy companies all round the world jump on the bandwagon? The big problem is that there is no useful international regulation. We will dig into these issues in Chapter 7 and Chapter 8.

But first... why should we care? The next few chapters look at the importance of the sky for ordinary people, for culture and civilisation, and for science and technology. Then we will get back to what we can do.

III. THE ROOF OF NATURE

September 25th 2020

IT'S A PERFECT DAY IN SCALPSIE BAY. Not a cloud in the sky. Just the slightest breeze. We are swimming in the Sound of Bute, looking across to the mountains of Arran. To our back is sand and rock and the marram grass dunes. We spot a seal, thirty yards ahead, bobbing up and and down, while our own wee dog is going crazy on the shore, because we have wandered too far away. Behind her are the dunes, a road, and then waving fields of corn. The sky is like a blue dome over the world.

A few hours later we are at our friends' house, drinking beer and enjoying the smells of dinner on the way. It's beginning to get dark, and the stars are coming out. Our friends are not stargazers - far from it - but like everybody they appreciate the beauty of a starry night. Its all of a piece with the natural pleasures of the world - the sun, the trees, stirring birds, night falling and the twinkling stars. Its all free and good, a kind of human birthright. These things belong to everybody - rich or poor, Scot, Bantu, or Inuit.

YOU DON'T NEED TO BE a serious stargazer or indeed

understand anything about the stars to enjoy these things, any more than I need to go about cataloguing butterflies to get pleasure from Nature. However, if you do have a little knowledge it only deepens your pleasure. For some, this is about the distinguishing and naming of things. Look, diving over there, that's a gannet, but over there that's a guillemot. See up there, that's Orion, and that bright red star is Betelgeuse. This way, we feel that the birds and the stars are a meaningful part of our lives.

Over the centuries, through slow and difficult observation and argumentation, humanity has understood more about stars (and gannets). This can deepen your appreciation further. That red star is actually a giant burning ball of gas, like the Sun but much much further away. There are red stars and blue stars and yellow stars - some are hotter than others, and they are very different sizes - giants and dwarfs. Why is that, and how do they burn? They are like enormous continuous Hydrogen bombs. What a thought, staring at that pinprick of light, that it shines like billions of bombs exploding every second. But now, see that bright star there - it's not in the same place as last month. It's not a star, it's a planet - a globe of rock much like the Earth, and like us, circling around the Sun. And now, see that faint star just over there, that you can only just see out of the corner of your eye? We know that star has its own planets, circling around. Imagine that.

I am a bit of a nerd of course, so I get pleasure from the intellectual insight in these things, and I am a professional astronomer, so my job is pushing that observation and argumentation further. But I am also a normal human being, and I love the buzz of simply struggling to visualise, and physically comprehend, these amazing things. Sometimes it's

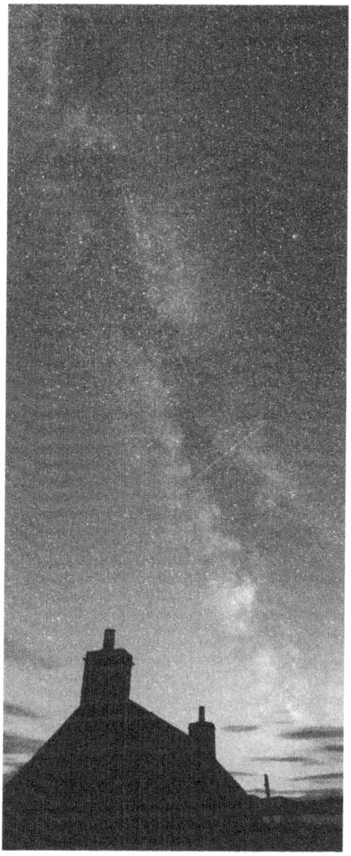

Figure 3.1: The Milky Way from Caithness. Picture by Gordon Mackie.

about feeling a sense of place, of being physically located. One of the best things about working at the Royal Observatory in Edinburgh, is that everday as I start my journey home from Blackford Hill, I can look around and understand where I am - there is Arthurs Seat, there is the Castle, and there are the Pentland Hills. In just the same way, on a dark night, I can stare at the Milky Way, angling across the sky, and feel in my bones that I live inside a giant disc of stars. Looking in this direction or that, out of the disc, I see the odd scattered star. Looking anywhere in this strip of directions, the distant stars blur together in their teeming millions, making the fuzzy band we call the Milky Way. This is where I live.

Then what about over there in Andromeda, that very faint fuzzy patch a few moons across, that I can only just see? It's another giant swirling disc of stars, completely outside the one that I live in. It's the most distant thing that you can see with your naked eye. With binoculars or a small telescope you can see many more of these swirling star cities, or galaxies. They are the building blocks of the Universe. We live in just one, that we call the Milky Way. The Sun is just one of four hundred billion stars in the Milky Way. We live on a rock that circles around that star. Our rock swarms with life, warmed by our Sun. And here we are. Picture that.

OF COURSE FOR MANY OF US, we get our buzz, our sense of wonder of the natural world we live in, from TV or from roaming the Web. David Attenborough, or Brian Cox, can, it seems, show us more amazing things than we would otherwise have dreamed possible. Those BBC camera crews can capture flight in slow motion, or wait for hours to see the young emerge from their desert burrows. Likewise we can see the latest pictures from the Hubble Space Telescope, or the computer animation made by Carlos Frenk's latest mega-simulation. This is all good. Its even better if we can link that TV show to our own physical experience; but at least we can all share in the wonder of Nature, above and below our heads. In Astronomy, just as in the study of wildlife, there is also a very important world in between the mass of citizens on the one hand, and the billion dollar space mission on the other: the passionate amateur.

These days, with good modern equipment costing just hundreds of pounds - the price of a new phone - amateur astronomers all over the world capture by far the best photographs of the sky. To me, when doing my day-job, I just want the numbers. I don't care much if its pretty. Amateurs often do quantitative science too; but they also care about creating the most perfect, most beautiful, most humanly meaningful images of what they see in the heavens. In Figure 3.2 you can see the Andromeda Nebula, our neighbour star city, captured lovingly by Mark Phillips from Edinburgh.

Figure 3.2: The Andromeda Nebula, photographed by Edinburgh amateur Mark Phillips. Not bad from the city that used to be known as "Old Smokey". ```

TODAY, THE NATURAL WORLD, our human birthright, the world we are part of, is gradually vanishing before our eyes. I see the loss of the sky as part of that environmental catastrophe. How many people have seen the Milky Way, and understand where they live? How many people have seen the Andromeda Galaxy for real - with their eyes, or with binoculars - or have any chance of doing so without travelling hundreds of miles? How many people see the stars at all? We are alienated from the sky. Yes, climate change, and the loss of biodiversity, is worse, far worse. But I see it as all one thing. The sky is the Roof of Nature. The loss of the sky is one more damned thing, the final straw. Of course, we can try to protect some areas of the Earth at least. But for plants, animals, and stars, this may not be enough; the world is connected.

That day in Scalpsie Bay, we walked along the shore

picking up pieces of plastic. It was some kind of official beach clear-up weekend I think. On Bute, this problem is not so bad as it is back in my Kentish seaside home town, but the problem is everywhere. When you drop a plastic bottle, it may not stay on your local beach, but will drift on the currents and could potentially wash up anywhere. Likewise, the microfibres you wash out of your clothes and flush away, will spread across the world. Drop a bag in Sydney and kill a fish in Mauritius. The plastic collects in horrific concentrations in the ocean gyres. The new sky pollution - swarms of satellites - is even more global. When you a launch a satellite in Low Earth Orbit, within an hour and a half it has circled the globe. You may launch from Florida or Kazakhstan, but you are instantly polluting Namibia and France.

IV. THE STARS IN THE HUMAN STORY

November 3rd 2020

IT IS ELECTION NIGHT IN THE USA, and tensions are running high all over the world. You can probably guess which way I sway. But for liberal intellectuals like myself, who have spent so many years feeling that we try to speak up for the ordinary folk against the rich and the powerful, it is uncomfortable to see populist politicians generate such adulation and acclaim in millions of those ordinary people. I find myself thinking of the Roman patricians Cassius and Brutus in the opening scene of Julius Caesar. As they lean over the bridge watching the gushing Tiber, Cassius talks bitterly of how the crowds shouted Caesar's name, and begged him to take the crown. He urges Brutus not to accept Fate meekly. They must stand against the tyrant. As he tries to stiffen his spine, what metaphor does Cassius turn to? He says:

The fault, dear Brutus, lies not in our
stars, but in ourselves that we are underlings

I love that quote. It sparks thoughts of power, action, and the stars, tied up in a perfect knot of language. Many of my professional colleagues would complain that the metaphor is not about astronomy at all, but about astrology, which many astronomers love to hate. I'm not one of them. Astrology was an early type of science, looking for patterns and predictability in the Cosmos. We may no longer believe that lights in the sky determine our characters, our fates, and our daily fortunes, but such ideas were not superstition - just science that has since been superseded.

The urge to understand, predict and control the world is deep. Likewise, the human experience of the night sky is so profound, and the mysteries of its patterns and regularities so intriguing, that it was inevitable that people wanted to find power and insight in these mysteries. Throughout history, the stars have been feeding our imaginations, and driving our mystical thoughts and reveries.

THAT GRIP ON OUR IMAGINATIONS has been there since the very earliest times. Some researchers have argued that depictions of animals in prehistoric cave paintings actually represent constellations, and were used by ancient peoples to keep track of time and the changing seasons, as much as 40,000 years ago. In prehistory and early recorded history, it seems that the practical importance and mystical significance of observing the sky were hard to separate. Keeping track of time and the seasons must have been crucial to the agricultural revolution - we needed to know when to sow and when to reap, and how to plan ahead for these things. The success of crops was crucial to our survival, and the ability to understand, calculate, and predict, a key feature in the social changes that came along, first with agriculture and then with settled cities. Our modern imaginations are fired up by theories and stories of prehistoric solar alignments, powerful priestly castes, and predictions of eclipses cementing the priestly hold over the panicking masses. Of course so much of this is guesswork and fantasy; we may never really know the detailed truth. But the deep importance of the sky is clear.

Meanwhile, the heavens became populated with depictions of our gods. Again, much is mystery. It is hard to tell how much of what we saw in the skies came from a love of telling stories, and how much from a desperate urge to find meaning. But somehow, every culture we know - ancient Greeks, Australian Aborigines, the Inuit, or the Polynesians - did much the same thing. The stories that go with the sky are imprinted firmly on our psyches. Ever since, through to modern times, images, metaphors and similes connected with astronomy and the heavens are sprinkled all over our story-telling and our art - from Giotto's exquisite painting of what we later came to call Halley's Comet, through the bitterness of Cassius, to Walt Whitman tiring of "The Learn'd Astronomer", and on to the stately march of Jupiter in Holst's Planets Suite. You may object that most of these things are really stories of the gods, and divination, rather than referring to the sky itself - but for thousands of years the sky and its stories were tangled up tight.

THE NATURAL WORLD CAN BE HOSTILE and frightening, as well as inspiring and comforting. The desire to predict and control is of course not just about the sky and the seasons. In ancient civilisation, the pattern of the yarrow stalks cast upon the ground was as important to study as the motion of the planets through the pattern of fixed stars. Unexpected events - a comet, an eclipse, a new star in the heavens - must have seemed particularly portentous. At heart, the urge to find pattern, meaning, and predictability is no different from the drive behind science. What changed was the growth of a philosophical and emotional desire to pursue the understanding of Nature as an end in itself, rather than just as the route to divination and control.

Figure 4.1: Modern Indian Astronomer Rathnasree Nandivada showing me the eighteenth century Jantar Mantar observatory in Delhi. I visited in 2014.

Gradually the study of the sky became professionalised, esoteric, and demanding specialised equipment in dedicated observatories - giant transit circles, clocks, and graded arcs. In early modern times, at the observatories of Ulugh Beg in Samarkand, of Tycho Brahe in Hven, or Jantar Mantar in Jaipur, the astronomical and the astrological purposes are quite hard to disentangle. But the giant structures and careful observation also had a very practical purpose - our mastery of Time.

AS CIVILISATION PROGRESSED, it became more and more important to track and forecast the passage of time. At first this meant the Calendar - forecasting the seasons, the dates of holy days and festivals, the dates of spring and neap tides, and so on. Observation of the sky remained important as astronomers realised that calendar and seasons were slowly drifting out of sync. Gradually humanity also became obsessed with the precise time inside the day - to predict sunset and sunrise, the timing of high and low tides, or the arrangement of meetings. The solution was partly mechanical - the development of better and better clocks. But the best, the most reliable, clock was the passage of the stars across the sky, as the Earth rotated.

Figure 4.2: Calton Hill in Edinburgh, showing the Time Ball on the Nelson Monument, next door to the Old Royal Observatory. Astronomers would observe the sky, adjust their clocks, and drop the Time Ball at one o'clock as a Time signal to sailors in the Firth of Forth. (Picture by Saffron Blaze, licensed CC BY 3.0)

The core job of state-funded astronomers was to maintain a master clock synchronised with the motion of the stars. Here in Edinburgh, Victorian citizens and mariners from Leith would climb up Calton Hill and peer through the window at the Observatory Clock to correct their timepieces, until Charles Piazzi Smyth began the Time Ball and the One O'Clock Gun to save them the climb. Originally time was always local, but the connection of cities by rail, and navigation over the global seas, meant that we needed a precise Universal Time - so that everybody could refer their clocks to the time at some special chosen point, such as Greenwich. Certainly in Britain, observation of the sky was closely tied to its industrial development, and its imperialist ambitions.

During the nineteenth and twentieth centuries, the mastery of Time continued to be central to the development of civilisation, but the role of observation of the night sky in that mastery declined. At the same time astronomy developed a much broader significance in the story of science and technology. That's the topic of the next chapter. In this chapter, what I have tried to do is to show that, far from being an esoteric and irrelevant pursuit, the study of the sky and stars has been a core part of human culture, religion and industrial development. Can we imagine history without the stars? Suppose the skies had always been hidden behind cloud, or hidden behind some glare of light, or utterly confused by a swarm of moving lights? I am sure we would still have had religions and cultures; just not the ones we know and love. But without the sky it is genuinely hard to imagine how we would have achieved an accurate mastery of Time, and all the practical civilisation that has emerged from that. Without the one clock we knew that we could trust - the rotating sky - what would we have done?

V. SCIENCE, TECHNOLOGY AND THE COSMOS

October 6th 2020

TWITTER AND EMAIL ARE BUZZING TODAY. The 2020 Nobel prize for Physics has been announced, for work on black holes. My colleagues are spreading their excitement and their pleasure, and swapping stories about the three very different and fascinating characters who are sharing the prize - Roger Penrose, Reinhard Genzel, and Andrea Ghez. I am recalling a panel discussion I was on a few weeks ago, involving a stimulating mixture of scientists, social scientists and artists, in which I described myself as "a black hole user". The discussion followed a lecture by Peter Galison, who fuses Physics, History, Philosophy and Art in his own work, and was part of the large team who produced the first black hole image in 2019. Black Holes capture the public imagination to such an extent that sometimes they feel like a kind of Hollywood entertainment. They seem a perfect example of human thought pushed to the limits - intellectually intriguing but abstract, beautiful but useless. This has some truth, but misses the real story. Black hole theory is indeed an abstract mental construction; but its study is at the core of the hard nosed science and technology which underpins the modern world. The work of Penrose, Genzel and Ghez is part of this story.

PENROSE THOUGHT DEEPLY FOR YEARS about the curvature of space, the principle of cosmic censorship, and whether our concepts of space and time still make sense when pushed to the limit of infinite density. These are fundamental and universal matters, but they were not developed in isolation. They were motivated by worrying about the stars. What happens when a star runs out of fuel to burn, and collapses under its own weight? What will it look like? Will something stop the collapse? These late twentieth century obsessions emerged from a long and tangled history of the interplay of physics and astronomy. Newton developed gravity to explain the motion of the Moon. Eddington tested Einstein's new theory of gravity by carefully observing the precise positions of stars during a solar eclipse. Key to the development of atomic physics was explaining the strange pattern of lines seen in the spectra of stars by Balmer. New elements were discovered in the stars. Worrying about how stars burn, and why they come in different sizes and colours, was crucial to understanding nuclear physics, which led in turn to the atomic bomb. The leader of the wartime bomb project, Robert Oppenheimer, worried in turn about the end of a star's life and how, when it finally collapses, its particles squash together to make a superdense neutron star. But if a star is too big, not even the neutrons can resist gravity. What then? A black hole.

If we had lived on Venus, with its thick layer of clouds, and had never seen the stars, how much harder would it have been to understand how matter and energy worked?

But, these strange ideas of neutron stars and black holes - are they real or some kind of elaborate mental fantasy? For decades astronomers searched for evidence in the real world - and found it. Neutron stars are seen by radio astronomers as pulsars; black holes as bright sources of X-ray emission, caused by surrounding gas heating up as it spirals slowly into the black hole. Like almost every other astronomical advance, these discoveries depended on new technology - huge radio dishes, interferometer arrays, rocket launches, spacecraft attitude control, new types of detector. The work of Ghez and Genzel is a marvelous example.

FIGURE 5.1 SHOWS THE TRACKS OF STARS in the centre of our own Milky Way, carefully tracked by Ghez and her team over many years, and also tracked independently by Genzel and his team. The tracks are the classic elliptical shapes expected for objects orbiting something massive, under the influence of the gravity of that central massive object. From the orbits, you can see where the massive object is, and work out its mass - several million times the mass of a star - but there is nothing there to be seen. This work has been a classic piece of slow, patient, rigorous science. But it would have been impossible without some impressive engineering and technology. The observations were made with huge telescopes in large buildings on distant mountaintops; challenging construction projects in themselves. The measurements were made using infrared detectors, a technology first developed by the military, but pushed hard and developed further by astronomers. The images were exquisitely sharp, a quality made possible by adaptive optics - again originally a military technology, but pushed and developed by astronomers, and deployed in laser eye-surgery as well scientific measurements.

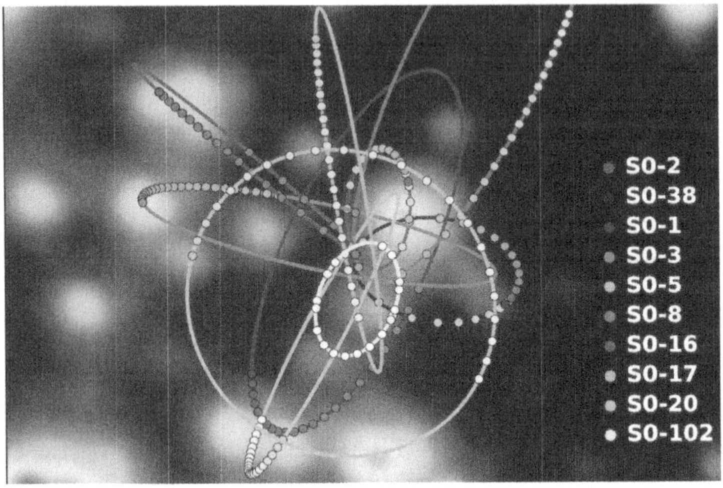

Figure 5.1: The tracks of stars in the centre of the Milky Way between 1995 anbd 2016. The stars appear to be orbiting around a massive central object - from which we see no light. These images/animations were created by Prof. Andrea Ghez and her research team at UCLA and are from data sets obtained with the W. M. Keck Telescopes.

Astronomy is not an ivory tower pursuit. It is part of the complex interacting organism of science and technology, and their manifestation in the industrial, commercial and political worlds. Sometimes astronomy benefits from the latest technology as an early adopter - for example infrared detectors, CCD cameras, the World Wide Web, phased arrays. In most cases astronomers also push industry by being demanding and high profile customers, carrying tax dollars in their pockets. Sometimes astronomy invents technologies which then move out into the world - for example the wifi connections first developed by radio astronomers. So modern society both loves and needs astronomy, just like it needs science overall. But pollution of the sky is putting it in danger.

THE OCCASIONAL SATELLITE STREAK crossing an astronomical image has always been a minor problem, but it has become steadily worse, rapidly so in the last two years, and has every prospect of becoming alarmingly worse. An example is shown in Figure 5.2. This is a so-called Wide-Field image, about the equivalent of five Moons across. Low-orbit satellites move across such an image in a few seconds, and so are only on each camera pixel for a very short time, but they are so bright that the detector tracks them all the way across, making a streak. Whats more, to get the deep pictures astronomers need, the integration times can be several minutes or more - during which time more and more satellites cross the field.

Such wide-angle imaging has become the norm in recent years. Its the best way to cover large amounts of sky quickly to make deep maps, but astronomers also survey the sky repeatedly to spot things that change - for example supernovae, flaring stars, and stars being ripped apart by supermassive black holes - rare but important phenomena. As I write this, I have been swapping emails with colleagues because we just spotted a flare up in the centre of a distant galaxy, and we want to catch this object with as many telescopes as we can before it fades. A streak in the wrong place, and we would never have seen it. As well as a scientific issue, this is potentially a planetary protection issue. Monitoring with wide field cameras to see what changes or moves is also how we spot Near Earth Objects - rocks that might crash into the Earth.

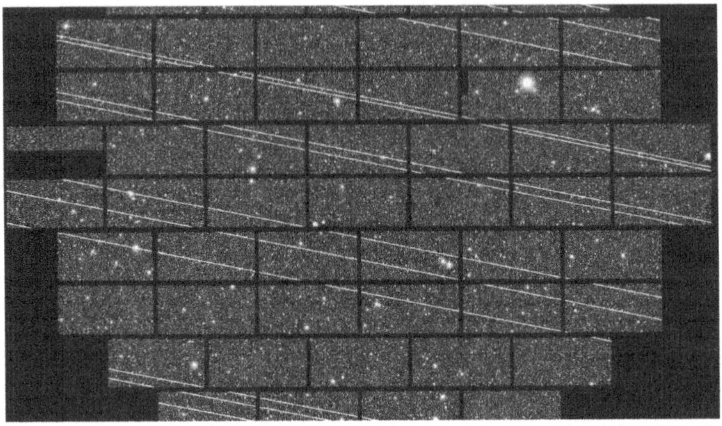

Figure 5.2: An An image of the sky taken by the Dark Energy Survey Camera in 2019. The effect in this case is particularly severe because many Starlink satellites are clumped together during the bright orbit raising phase; however, soon this will be happening evry few weeks. Credit: CTIO/ NOIRLab/ NSF/ AURA/ DECam DELVE Survey.

Wide-field imaging is particularly susceptible to satellites - there will nearly always be multiple satellites streaking across during the exposure. Narrow-angle imaging is also at risk in a different way. You might be lucky - but if you are not lucky, you can write off that entire observation.

AND WHAT ABOUT RADIO TELESCOPES? Aren't they safe? No such luck. Starlink satellites beam their data down to Earth by radio signals around 11GHz, where many observations of great scientific importance are made. What's worse, its not just a question of annoying streaks through the data... satellite constellations will produce a bright glare affecting every observation around this frequency.

The way radio telescopes work is much the same as your TV aerial. You dangle a piece of wire in the air, and incoming radio waves excite the electrons in the wire, which makes a tiny oscillating voltage which we can amplify with clever electronics. Some radio telescopes have dishes, but they are mostly just big concentrators, with an antenna in the middle. The problem with antennas is that they are potentially sensitive to radio waves coming in from any direction at all. The antenna shapes and the concentrator dishes help to make to make the system more sensitive in the directions they are pointed, and you can combine the signals from an array of antennas to sharpen the directional sensitivity even more. You can never completely get rid of the problem, but these various techniques can make you thousands of times less sensitive to signals coming in from the side. Unfortunately Starlink satellites are expected to be billions of times brighter than the typical cosmic radio sources we want to look at. It's as if you are straining at night to see someone with a dim flashlight miles away on a distant hillside, but you are standing on the side of a busy road, and all you can see is the glare of headlights in the corner of your eye.

Radio telescopes, like optical telescopes, have a dark-sky problem. Rather than the glow from city lights scattered back by the atmosphere, their problem is radio interference from just about any electrical device. The ignition of a car starting can be seen many miles away. So radio telescopes, like optical telescopes, are located (expensively!) in remote places. But just as with optical observing, there is no escaping the satellites, no matter where you are located. Radio astronomers are well used to seeing communication satellites - indeed Jodrell Bank not only noticed Sputnik straight away in 1957, but using an attached radar set, even tracked the rocket that launched it. However until now, most communication satellites have been in very high, geo-synchronous, orbits - they are in predictable positions and move very little, so can be kept out of the telescope "side-lobes". However, for the new low-orbit constellations, there will be hundreds in the observable sky at any one time, sweeping across the sky in an effectively unpredictable manner.

Right now, the international community is clubbing together to build the Square Kilometer Array, the most powerful radio telescope system ever built, aimed at elucidating the origin of structure in our Universe and many other problems. As I write, I see they that have just issued a press release reporting the results of their recent study of the satellite problem. The complete Starlink constellation will produce an effective constant background noise that means that many observations will take 70% longer to accomplish, effectively wasting billions of dollars of taxpayers money from countries all over the world. Occasional fly-bys in just the wrong place could produce a signal bright enough to burn out some of the very sensitive pieces of electronic equipment, designed to detect faint signals from the edge of the Universe.

SO MUCH FOR RADIO ASTRONOMY. What about space astronomy - won't that be safe? No such luck. Telescopes in space can just as easily have their observations ruined by other spacecraft. The Hubble Space Telescope is at a height of 540km, somewhat below the 600km of initial Starlink launches. But that means a satellite can actually pass very close… Figure 5.3 shows an observation made with the Hubble Space Telescope in November 2020, showing a bright fat streak. According to Jonathan McDowell, the orbit data he has for SpaceX satellite Starlink 1619 suggests that it passed within 80km of the Hubble Space Telescope. At this distance, the satellite would have been very bright indeed compared to the faint things the telescope was trying to capture, and out of focus; the result is not just a thin streak but a very fat one.

Figure 5.3: An observation made using the Hubble Space Telescope in November 2020. The streak seems likely to have been made by Starlink 1619. (Image credit: Mikulski Archive for Space Telescopes (MAST). Science PI: Simon Porter.)

Some astronomical observatories are at much more distant locations, very far from Earth, but spacecraft used by our sibling subjects such as space weather studies and Earth observation are routinely in low orbits, and look down and sideways as well as up.

Furthermore, Low Earth Orbit is looking like an ever more dangerous and crowded place to be. Will low orbit space become so crowded that collisions will become inevitable? This is a complex problem, but the short answer is yes. Satellites pass within a few km of each other every day. In February 2009, Iridium 33 collided with Kosmos-2251, creating an estimated thousand pieces of debris. In September 2019, the European Space Agency announced that it had to move its Aeolus lidar satellite, studying atmospheric dynamics, to avoid a collision with Starlink 44. There is an international system to track satellites and so try to avoid such collisions; in this case, mistakes were made and lessons learned, but the point is that any system is imperfect, and more collisions will happen as space becomes more crowded.

And it's not just collisions. Every launch leaves behind rocket stages, nose cones and so forth. As satellites complete their missions, they become uncontrollable pieces of metal. Worse, it seems that a large fraction of debris items arise from explosions of satellites caused by unspent fuel. In low orbits we can also add rocket debris and failed payloads. In 2007, China tested an anti-satellite weapon on one of its own satellites, making thousands more pieces of debris.

Tracking and controlling the major active satellites is one thing, but tracking and avoiding spacecraft debris is far harder. The number of pieces of debris currently outnumbers the active satellites by an order of magnitude, and thats just the number of objects larger than a few cm, which we have some chance of cataloguing. Small pieces of debris, even floating flecks of paint, now routinely cause ablation and pitting in active spacecraft, which need shielding from such impacts. The problem has become worse than that caused by the existing natural population of micro-meteroids. The wallop with which these things hit is amazing; in a 500km orbit, objects are moving at 7.6km/sec. A counter-rotating speck of iron only half a millimetre across has about the same kinetic energy as a cricket ball delivered at 100 km/hour. A piece of iron 2cm across has about the same kinetic energy as two medium sized cars in a head-on collision at 50 miles an hour.

Figure 5.4: A hole made in the first generation solar arrays on the Hubble Space Telescope by a small piece of orbital debris. The picture was taken by Mark McCaughrean of the European Space Agency in his own office, complete with his reflection - a kind of space-selfie. Image credit: NASA/ESA/M.McCaughrean.

SO WHAT CAN BE DONE? Its worth noting already that SpaceX have already engaged with astronomers to discuss mitigation possibilities, and OneWeb have made nice public statements about "Responsible Space", but we need to ask whether this is enough. We will tackle this problem again in Chapter 8, after looking at just how bad things could get in Chapter 7. But let's take a first quick look, and knock some simple ideas on the head.

Why don't we just start doing astronomy in space? Well... nice idea, but who is going for pay for that? Some things can be done better by observing from space - and we already do so - but even then it is much more expensive, and is only done when we can make the strongest case that it is really necessary. Other things are still better from ground, mostly because very large telescopes are needed. The Hubble Space Telescope is two metres across. The new "Extremely Large Telescope" being built by the European Southern Observatory is forty metres across. Thats basically impossible to fly. Almost of all modern radio astronomy is impossible to do from space, using connected arrays of large dishes. Finally of course there is the debris problem, which may well make all of space science, let alone cosmic astronomy, very much harder to do.

Next, we hear it said that the satellites are only visible for a brief period in twilight. As we discussed in Chapter 2, this is an oversimplification. At higher observing latitudes, the visibility is much longer. Slightly higher satellites, like those planned by OneWeb, are visible all night. Many of the most important scientific observations need to be done at the beginning and end of the night - for example catching the Near Earth Orbit "killer rocks" - or have to be done at a very specific time that we can't change - for example observing the predicted transit of an exoplanet across the face of its parent star. Then finally, for radio astronomy, there is no night and day. The satellites don't just shine in reflected light - they are beaming down their glare of radio signal all the time.

Should the satellite operators paint key parts of their satellites black? Or put shields on them that reduce reflection towards the ground? Following discussion with astronomers, SpaceX have indeed done experiments of this kind, and it does help somewhat; the brightness we observe seems to reduce by a factor of two or three. This is at least promising for reducing the problem for naked eye observing, but nowhere near enough to remove the problem for large professional telescopes, or even observations by keen amateur astronomers with small telescopes and binoculars.

Should the satellite operators publish their satellite orbit data so that astronomers can predict where they will be when? For radio astronomy this doesn't help at all. For optical astronomy it might be feasible, but its hard to do, and writes off hours of (expensive) telescope time which can't then be used. Maybe astronomers should adapt their data analysis software to deal with the streaks in the data? This is partly plausible, but at the cost of probably hundreds of person-years worth of software engineering time around the world, and it makes the life of every ordinary astronomy graduate student harder than it needs to be. Furthermore such software can never recover the obliterated data - instead, there are areas of sky that would need to be re-observed, again effectively wasting huge numbers of taxpayer dollars.

WE HAVE ARRIVED AT A STRANGE PARADOX. Earlier in this chapter I made the point that Astronomy is not a monastic pursuit, but rather an integral part of the modern story of science, technology, commerce, and innovation. Yet here I am complaining that some particular beautiful and visionary technology innovations are a bad thing. How do I square this apparent contradiction? That is the subject of our next chapter.

VI. LIVING IN THE HEART OF THE BEAST

October 19th 2020

THE MODERN WORLD IS A GRIMLY FASCINATING blend of pride and guilt; of creation and decay; of complexity and chaos; of beauty and violence. In each case it feels like we can't have the Yin without the Yang, the Light without the Dark.

AS I SURFACE THIS MORNING FROM MY SLUMBERS, a voice from hundreds of miles away emerges from a box at my ear, bringing me news of the world, and playing music for my delight. I cut open a new packet of muesli, add defrosted frozen berries, and pour on soymilk. Yesterday, all those things were magically waiting for me at the store. Out on the road, at the turn of a key, I control a machine constructed in Japan, which takes me to my place of work twenty times faster than my legs will go. On top of Blackford Hill it is almost achingly beautiful, staring out at both Nature and Civilisation - the Pentland Hills, the Castle, and the suburban sprawl. At the Observatory, in both carved Victorian stone and shiny glass and concrete, we meditate on the depths of the Universe, but we also cut metal, manage huge projects, and teach physics to a bright and keen new generation. These are the days of miracle and wonder.

But the radio brings news of war, poverty, and racial tension.The soybean monoculture, it seems, is bad for the planet. My car is burning fossil fuels, adding greenhouses gases to the warming atmosphere. The weather seems more freakish every year. My breakfast, so conveniently fetched from the bottom of the road, has actually been assembled from across the globe, burning more gases, exploiting distant communities, and creating yet more plastic. There is plastic blowing across the park on my drive to work. The world is not all architectural splendour; it is also rubbish piles, pollution, and industrial wasteland.

The wonders and the rubbish, the gleaming cities and the industrial blight, come as a package - or do they? This is perhaps the central dilemma of modern humanity - can we build an affluent and technological world without polluting it? The satellite constellation issue seems such a good example. "You love the Internet? You want movies on demand? Well, this is what we need to do. I am sorry, you can't have it both ways." Sometimes in Astronomy we like to see ourselves as pure and selfless, pursuing knowledge for its own sake, above the grubby business of making money and political horse trading. But the truth is very different.

We are not Luddites, but quite the opposite. As we saw in Chapter 5, astronomers eagerly consume and create technology, and run massive engineering projects to achieve their ends. To do this we work with industry, and negotiate with government. Why does almost every government feed large amounts of precious public money into science, including astronomy? Because they understand that the search for knowledge, the invention of technology, and the spirit of enterprise, are joined together in a complex interacting whole that needs to be fed and kept healthy. It is a lumbering but powerful beast, and we live in its heart.

If now, we get distressed, as our pristine cosmic visions become polluted, it does not befit us to take some kind of moral high ground. We have been part of the story of the unrestrained ambition of the modern world, and we are slowly starting to look around and understand what we have done. Let me give you some examples.

ASTRONOMERS SHOULD BE ESPECIALLY AWARE of the crucial importance of greenhouse gases and runaway heating - the surface of cloudy Venus is at an unliveable 460 degrees Celsius. In the computer models we build of stars, we can see how the temperature structure of those stars depends on their opacity, which depends in turn on the elements that build up within them. We understand the physics - it's not that hard. But for decades, we have been eagerly flying around the world, helping to dump carbon into the atmosphere of our own planet. We have telescopes around the globe and travel to use them a few nights at a time. We fly constantly to project meetings and committees, and several times a year to international conferences. We love to meet, talk, and debate. There are astronomers everywhere, and we feel like citizens of the world. As we greet our friends and colleagues from Brazil and China and France, and sample other cultures and foods, it feeds our comfortable sense of being liberal paragons. (Apologies to my conservative friends, who also feel like citizens of the world.)

I just wrote the above paragraph in the present tense, but

in fact things have already changed. Our habits have been altering fast, pushed partly by guilt over our contribution to climate change, and partly of course by the COVID-19 pandemic, which has shut down air travel almost overnight. Now, instead of spending half my life in hotel rooms and airports, I spend half my life on Zoom or Teams. There are many things that are positively better about this new online world of scientific discourse and project management - less time wasted, less jet lag, the enthusiastic chat windows, the ease of making a cup of coffee while carrying on listening. I am sure we will still sometimes join our world-wide friends in the flesh, and drink beer at the same time as arguing about quasars, but it will become a rare treat. Cutting down our travel was the right thing to do, and we have found that it can be done.

WELL, WE MIGHT SAY, at least we are not spraying the land with DDT, cutting down the rainforest to plant soybeans for cattle feed, or displacing communities to dig for oil. True, but we have blasted the tops off mountains and constructed concrete and metal monsters in pristine wilderness, sometimes with insufficient care and thought about who or what else might have a stake in these beautiful places - the red squirrels on Mount Graham in Arizona, or the native people in Hawaii, who see Mauna Kea as a sacred place. The planned building of the Thirty Meter Telescope (TMT) on Mauna Kea has been a very public and scarring episode. Thc top of Mauna Kea is like a small town made of telescopes, but the plans for TMT led to an activist uprising, local people crying no more, enough! This isn't the place to revisit all the details and complexities of the arguments over TMT and Mauna Kea, but just to note that many astronomers were naively shocked, and it split the community.

Some said: "But we are the good guys! We are studying the Universe, for the benefit of all mankind, not for selfish greed. How can that be a problem?" Others said: "Oh. I must admit, this looks like colonial imperialism... An occupying power turns up and erects huge buildings that are nothing to do with the local people". I find myself swinging between these views - I really want the local people to see that studying the cosmos is a beautiful thing, good for us all, and join with us. But then I flip, and see that it must seem little different from Missionaries arriving waving a Bible, and saying that they bring civilisation, and redemption.

ONE MORE THING. It would be bad to give the impression that the Internet is just for shopping, or gaming, or for the split-second decisions of those greedy bankers. Astronomers, like other scientists, have been part of the drive for computer networking since the 1960s. In the 1980s, we were constantly emailing each other, years before my non-scientist friends had any idea this was possible - and also doing live split-screen "phone" conversations. In the 1990s, the Web was of course invented by our Particle Physics friends, but the European Southern Observatory had one of the first few working websites. Radio astronomers have always been obsessed with the highest possible bandwidth for data connections between their telescopes, and played a key role in the development of Wifi.

Starting in 2005, I helped to found an international effort

called "The Virtual Observatory". The idea was get all astronomical data online, and to make it easy to access and mix and match, wherever it lived, so that you could play with the world's data while sitting on your sofa. For this to work, we needed world-wide agreed standards - on data formats, access protocols, and so on. Like any other international agreements, this required much negotiation, compromise, and thrashing out of details, which tends to be best done over coffee or beer. So paradoxically, this drive for an online world involved me in more travel and meetings than ever, piling up that carbon guilt.

So astronomers want global Internet access as much as everybody else; and indeed part of the drive for the Virtual Observatory was that we would try to make doing astronomy as easy in a developing country as a rich one. One thing that became obvious was that end-users are quite impatient. If that online data analysis takes minutes rather than seconds, they lose confidence. But they don't yet insist on milli-second latency…

WHERE HAVE WE ENDED UP, and what has been my point in this chapter? In the eighteenth century, two contrasting images of the Scientific Revolution emerged, as exemplified in two famous paintings. Figure 6.1 shows a gathering of the Lunar Society of Birmingham - eighteenth century citizens keen to join in the new world of knowledge and science. From our modern standpoint, the idea of killing a bird in order to discuss science does not seem civilised at all, but at the time the image was clearly meant to exude the optimism of the new, rational and affluent world.

Figure 6.1: "An experiment on a bird in an air-pump", by Joseph Wright of Derby, 1768, showing various members of the Lunar Society of Birmingham. Sourced from wikipedia. The original is in The National Gallery, London.

Fig 6.2: Newton, by William Blake, 1695. Sourced from wikipedia. The original is in Tate Britain, London.

Figure 6.2 shows William Blake's vision of Isaac Newton, measuring out the world, reducing it to numbers. Some today might use this as a positive image, but it is certainly not what Blake intended - he wrote

> *Art is the Tree of Life. Science is the Tree of Death.*

For Blake, it was all part of the creation of those Dark Satanic Mills. And of course he was right. Science is completely tangled up in the Industrial Revolution and the creation of the modern world, for good and for ill.

For Astronomers, it is therefore certainly not a case of "science good, industry bad". We love technology, and the industrial development as well as the knowledge that makes it possible. But like everybody else dwelling inside the Leviathan of science, technology and industry, we are coming to grips with what it should mean to work and build responsibly. Sustainable and responsible science, technology and industry should be possible. In Chapter 8, we will try to look at how we can do that, dipping our toes into that ocean of complexity. But before then, as a final motivation, let us look at just how bad things could really get.

VII. PANDORA'S NIGHTMARE

October 23rd 2020

NEWS IS COMING IN FROM PLANET 4589. Seventy five satellites launched by SpaceX rivals OneWeb have been spotted raising their orbit heights, after a long quiet period during OneWeb's financial difficulties… Everyday, more news of what is going on in the space around the Earth reaches my computer in short packets of information from Planet 4589. Sadly, I don't have a secret interstellar communication channel. Planet 4589 is the Twitter handle of Jonathan McDowell. By day, Jonathan is a professional astrophysicist at the Harvard-Smithsonian Center for Astrophysics, researching the energy distributions of quasars, and working in the data centre for the Chandra X-ray Observatory. By night, he is a space hobbyist. Since his teenage years, he has been collecting and collating information on every spacecraft launch that has ever happened, until today his database and associated website is a valuable international resource. A handful of others, such as T.S.Kelso, have likewise obsessively tracked, analysed and discussed the world of rockets and satellites.

Figure 7.1: A glass photographic plate taken by the UK Schmidt telescope during the 1980s, held up in front of my office window by Catherine Heymans. You can see the Small Magellanic Cloud, several bright nearby stars, and thousands of faint distant stars. Afficionados will spot that the plate is upside down..

For McDowell, there has always been an intersection between his daytime and evening worlds - a tiny fraction of those satellites were astronomical observatories. But now these worlds are in collision, as the satellite constellations become a threat to astronomy, and McDowell analyses and simulates their impact in professional publications, and contributes to scientific working groups fretting over what to do. Like Jonathan, many astronomers round the world have a sense of sudden awakening, as we realise what has been happening, and how fast it could change, beyond our control. For years, we have been worried about light pollution, but at least we could retreat to our mountain fastnesses and keep our own dark skies, while being saddened that those beautiful skies were available to a smaller and smaller fraction of humanity. The problem of light pollution is like a steam roller - slowly but steadily flattening the world, while we try to hide away in ever smaller corners. For satellite constellations we need to reach for a different metaphor. Perhaps we are in a peaceful garden, when someone foolishly pokes an insect nest. Suddenly the air is thick with swarming angry wasps, and you cannot see how you can possibly get them back.

IN EDINBURGH WE STORE THOUSANDS of large glass plates, every one recording a patch of sky from the survey of the southern sky started in the 1970s by the UK Schmidt telescope, based in Australia. Each plate, a foot across, covers an area of sky equivalent to about 13 full moons side-by-side, in each direction. Every so often a passing satellite would cause a streak across the plate. This was annoying - ruining the measurement of that piece of sky - but it also had its own beauty. It was moving to think of a small machine that people had lifted into the sky, a link between people and the cosmos. And every satellite was doing something useful or interesting - mapping the Earth, beaming down TV, checking for unauthorised bomb tests, measuring the galaxies. Years later, it was even more fascinating to sit in your own backyard and watch the International Space Station drift gracefully across the sky. Like the occasional bee or wasp in our peaceful garden, the satellites seemed natural and useful. But now the swarm is being released and it all feels very different.

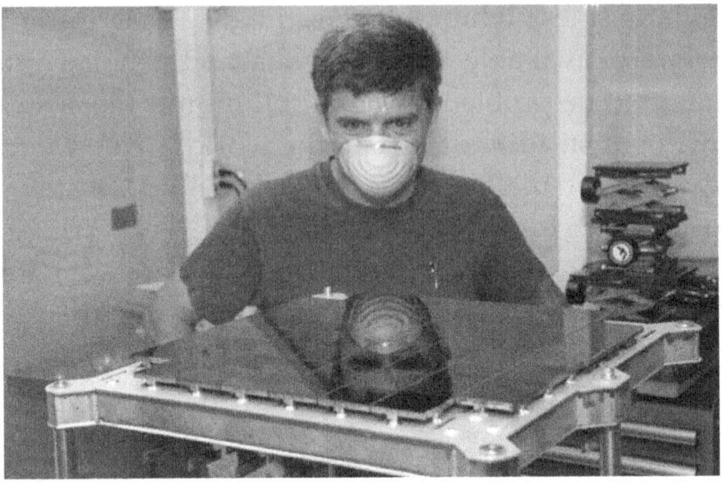

Figure 7.2: The mosaic of CCDs that make up the detecting device for the camera on the PanSTARRS-1 telescope in Hawaii. John Tonry is seen working on the device in the lab before it was inserted into the camera. The modern detector is much more efficient than the old photographic plate but about the same size.

As I write, about 800 SpaceX Starlink satellites have been launched, in three distinct waves. Meanwhile, they have approval by the US Federal Communications Commission (FCC) for 12,000 so far, and have stated plans for 42,000 satellites. But its not just SpaceX. Several other companies - OneWeb, Amazon, Boeing, Teledesic - have made known only slightly less ambitious plans. Earlier this year, Analytical Graphics, an American company, investigated filings made to the FCC - applications for potential satellites - and found 107,000 satellites proposed to be operational by 2029. Over the last few weeks, news from other countries is creeping out - 500 planned by a Russian company, over a thousand planned by three different companies in China. Nobody knows of course how many of these plans will turn into reality - but what we do know is that Pandora's box has been well and truly opened. How many satellites will there be in the sky in a few years? Fifty thousand? A hundred thousand? Three hundred thousand? We don't know, but any of these figures is plausible.

HOW MANY STARS CAN YOU SEE in the sky with your own unaided eyes? Most people, when asked this question, guess in the millions, because the sight of a star-spangled sky is so awe-inspiring. In fact, with the naked eye you can see just a few thousand stars. So will our hundred thousand or more satellites swamp the sky? Well, not quite. From any one spot on the ground you can't see all the satellites, as most of them are hidden behind the Earth. How many you can see depends on the height of the orbit. For satellites at 600km, you can see about 4% of them - but thats enough that potentially there will be similar numbers of stars and satellites.

Not all of those satellites will be bright enough to be visible to the naked eye, especially if some of them are in higher orbits. Since astronomers made their initial shock known, SpaceX have in fact worked on ways to make their satellites less bright, and it does make some difference, but of course we have no idea whether all those other companies will try as hard. Also, as we discussed in Chapter 2, low-orbit satellites will only be bright for parts of the night - although the parts when most people might be watching. The satellites are at their brightest just after they have been launched, and most of the spectacular streaks photographed so far have been in this satellite-raising phase. SpaceX plans include regular de-orbiting of existing satellites, and launching of new ones, with something like a turnover rate of 10% per year - so there will always be a population of easily seen satellites. Well anyway, we can see how hard it is to make clear predictions about the future, but it looks likely that not long from now there will be hundreds of easily visible satellites drifting across the sky, every evening and every morning.

Suppose now you are an amateur stargazer, with a small telescope, or perhaps gazing at the sky with a pair of binoculars on a camping holiday. With a 35mm lens in a good sky you can see many more stars, certainly outnumbering the satellites, but the satellites you will see will be a constant distraction. In a typical 7 degree field, if there a hundred thousand satelliltes in total orbiting the Earth, at any moment you will see about ten of them through your binoculars. As you watch, each one takes about 30 seconds to cross the field of view, and all the time, more and more of them are swimming in.

Imagine next taking an image with a large professional telescope. We can now see huge numbers of faint stars and galaxies, one of which may be just the one we want to measure. A wide field camera, like the one shown in Figure 7.2 above, may have a field of view a few degrees across, and will be used for stepping across the sky to make a kind of mosaic survey of the heavens, in a series of snapshots taking maybe thirty seconds each. A more typical camera might have a field of view only a tenth of a degree across, and is used to get the deepest pictures we can. To get these deep pictures, astronomers integrate - i.e. they take long exposures, collecting as much light as possible to see the faintest things. Individual exposures could be up to an hour, and sometimes many individual shots are co-added together.

Figure 7.3: A satellite-streak selfie, taken by Merlin Fischer-Levine. See text for detail.

You can see why these long exposures matter in Figure 7.3. Merlin Fisher-Levine was in Chile, working on testing the "Auxiliary Telescope" which sits next to the main telescope of the Rubin Observatory, a giant new survey telescope due to get underway in 2023. The Andean skies are beautiful, so Merlin put his phone on a tripod, stepped back and took a selfie, with his Android phone in astrophotography mode, with the Magellanic Clouds in view. During that two minutes, he was photobombed by a bright satellite. Whereas to the human eye, a satellite is a moving dot, in a long exposure it makes a streak across your image. The fuzzy patch to lower left is the Small Magellanic Cloud, the same dwarf galaxy we can see in Figure 7.1, the glass plate held up by Catherine Heymans.

The Rubin camera will have a field of view about three degrees across, and the Small Magellanic Cloud will fit neatly within it. Right now, only some survey snapshots will be photo-bombed, but once we have say a hundred thousand satellites, there will be something like ten or eleven streaks in every image. So what about those deep image cameras? Although their field of view is much smaller, over the space of an hour we can expect around 30 streaks in every image. There is no escape, and making the satellites a wee bit fainter makes very little difference. Those deep images are aimed at detecting very faint galaxies at the edge of the universe, planets circling around nearby stars, or faint moving smudges that actually may be rocks 100 metres across on a collision course with the Earth. To achieve this, the telescopes are large and the detectors very sensitive; the bright satellite streaks will always be seen, and in fact will often saturate the detector, making multiple false streaks through electronic cross-talk.

THINGS ARE EVEN WORSE - far worse - for radio astronomy. In Chapter 5 we described how radio antennas basically see the whole sky at once. The dishes they sit behind, the clever structure of the antennas, and phasing between related antennas, all help to concentrate nearly all their sensitivity into the direction they want to look, but their sensitivity to things "out of the corner of the eye" never goes to zero, especially in a few "sidelobe" directions here and there around the sky. With the handful of satellites in geostationary orbit, in well known and slowly changing positions, you can arrange your observations to keep those satellites out of the sidelobes. With a hundred thousand satellites in low orbits, there will always be hundreds or thousands of satellites all over the visible sky, shouting loudly in radio noise. They will be moving rapidly, and more and more will coming over the horizon all the time. Even if we knew all their orbits accurately enough to predict their positions on the sky second by second (an optimistic hope), it may be impossible to plan observations so that all the satellites can be avoided. The most sensitive radio telescope systems may become essentially unusable.

SO - IT LOOKS LIKE THINGS COULD GET REALLY BAD for visible light astronomy, and catastrophically bad for radio astronomy. Perhaps astronomy can retreat to space? Unfortunately, as we saw in Chapter 5, even space observatories can suffer from satellite pollution. Worse, the space around the Earth could become too dangerous to use. As we saw in Chapter 5, collisions do happen - not just calamitous crashes between two whole satellites, but constant damage by small pieces of debris. In the 1970s, when the satellite population was still modest, NASA scientist Don Kessler saw the problem getting worse, and predicted that the break-up of satellites would produce an unstoppable run-away, now known as the Kessler syndrome.

Why does this run-away happen? As objects collide or break up, the number of other objects you can collide with goes up. Of course, their cross-sectional area goes down - but at a slower rate than the number is increasing. Crudely speaking, this is because smaller objects have larger surface area per unit volume. (That's why ants get blown about in the wind and bears don't). So the collision rate goes up for the smaller objects. So you get more collisions, more debris, and the rate goes up again... and you end up with an exponential run-away.

Now it's actually a little more complicated than that of course. Satellites aren't spherical blobs, and their orbital motions are complicated and various... But the general idea remains. And indeed, it is quite possible that this kind of collision cascade has been involved in the evolution of the rings of Saturn, and the asteroid belt.

Figure 7.4 shows the growth over time of all space objects, as tracked by Jonathan McDowell in his "General Catalog of Space Objects". Note that "left overs" means rocket stages, nose cones etc that have been deliberately left behind, whereas "debris" means accidentally created pieces. Two things stand out. One is the big jumps caused by the Chinese weapons test in 2007, and the Iridium-Kosmos collision in 2009. Once collisions become more common, you can see how the run-away can happen. The second stand-out feature is the upturn in the total caused by the recent increase in launches, which is *not yet* reflected in the debris numbers. Not yet - but it surely will be.

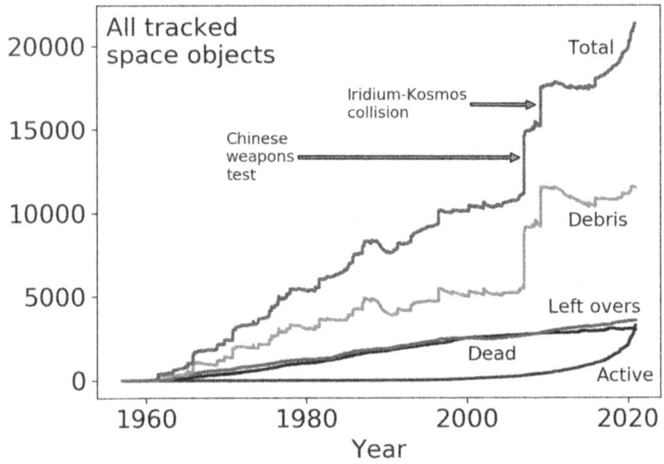

Figure 7.4: The growth of all tracked objects in space over time. Note that pieces smaller than a few cm are invisible, but there will be many more of them! Data provided by J.McDowell.

Opinions differ about how close we are to the Kessler syndrome; but what we do know is that a huge new population of Low Earth Orbit satellites will push us closer. Most of the live satellites and orbit debris are in either Low

Earth Orbit or Geosynchronous Orbit (see Figure 7.5), but the density in low orbits is far greater. In fact, a large fraction of launches are into polar orbits, and where these cross near the North and South Poles the density, and probability of collision, are greatest. Simulations made by the European Space Agency (ESA) suggest that even if we were to stop all launches now, we are already past a critical point; collisions and fragmentation and collisions will produce a cascade anyway - it's just that it will accelerate faster if we keep adding to the low orbit population. ESA have a program experimenting with actively removing dead satellites from orbit.

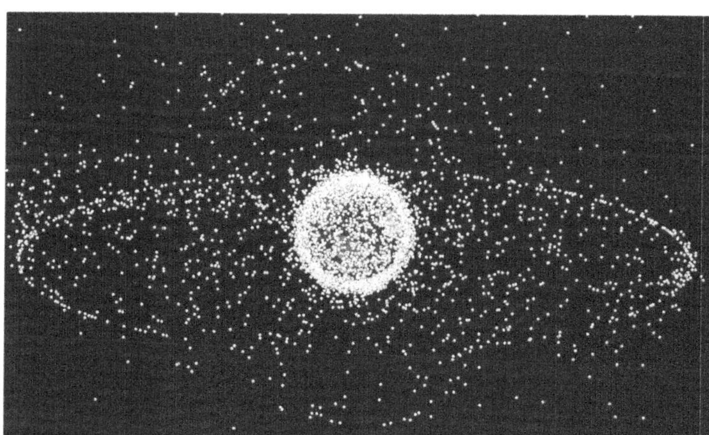

Fig 7.5: Locations of known space debris, from the NASA Orbital Debris Program Office. The outer ring is objects in Geo-synchronous orbit. The inner shell is objects in Low Earth Orbit.

ESA reports that it undertakes a handful of avoidance manoeuvres every year. However, this takes fuel or power. The dead satellites, left over parts, and pieces of rocket debris are all uncontrollable. To this list we can also add failed payloads. In the first two waves of Starlink launches, around 3% were failing. They now seem to have reduced the failure rate considerably - but who knows what the future holds, with dozens of cowboy companies?

The orbits of satellites in low orbits gradually decay due to atmospheric friction, and those satellites descend to Earth and burn up. However, this process can take several years. Space X plan to constantly decommission and "de-orbit" their satellites, with a turnover rate of about 10% per year. Given the expected rate at which satellites descend to Earth, there could be easily be a permanent population of a thousand or more dead satellites the size of a desk circling the globe.

ONE LAST SCARY AND DEPRESSING TOPIC for our survey of the dystopian future: advertising in space. We are already seeing product placement - the Pizza Hut logo on the side of a Proton rocket, and a Pepsi can carefully floated outside the Russian space station. Some will think this is ok, some not, but in any case it won't ruin our view of the night sky. What could is giant glowing billboards in space, visible from anywhere on Earth. As commercial spaceflight develops and launch costs come down, this gets ever more tempting. There have been several failed attempts - for example the 1989 French "Ring of Light" project, intended to celebrate 100 years of the Eiffel Tower; a proposal by Space Marketing Inc. to launch a space billboard in 1993; and a collaboration between PepsiCo and Russian company StartRocket aimed at launching a billboard in 2019. There has been one partly successful project - in 2018 a US/New Zealand company called Rocket Lab successfully launched a kind of 1 metre disco ball called "Humanity Star", but it was fainter than intended and only marginally visible to the human eye. Rocket Lab claimed it was "a bright symbol and reminder to all on Earth about our fragile place in the universe". Others called it "glittery space garbage".

Obtrusive advertising in orbit or on the Moon is going to happen unless the international community agrees to stop it. As well ruining the night sky, it is going to make the debris problem even worse…

STREAKS, RADIO GLARE, RAMPANT COLLISIONS, HORRIBLE ADVERTISING … it all seems such a grim future. What then must we do?

VIII. WHAT THEN MUST WE DO?

November 11th 2020

ANOTHER EMAIL NOTIFICATION is sliding across the top corner of my laptop screen. Roy Williams says "seen this, Andy?" and links to an electronic preprint of a new paper by Meredith Rawls and others, which analyses the assumption that the Starlink megaconstellation is needed to provide cheap internet access to poor parts of the world. They conclude that the business case for this is weak, as there are few places where this method of connection is both needed and affordable. It is interesting that Rawls *et al* have written an overtly critical article. Rawls works for the Rubin Observatory (as do Roy and myself, indirectly, as part of a UK collaboration). The Rubin project have done a good job of engaging with SpaceX, and the SpaceX scientists and engineers have responded well, developing improvements that might help. Meanhile some colleagues have been getting irate and starting petitions, while others are saying "don't rock the boat, we need the space industry", and yet others are beavering away very hard, forming international working groups to study the problem, and so on. What is the right approach? What then should we do?

GIVE UP AND ROLL OVER? Maybe we should just get real. This is the way of the world. Inexorable progress. Huge commercial forces. People want the Internet, or want faster Internet, wherever they are. A little sky pollution is not too high a price to pay to make so many people happy. Financial traders want the fastest connections they can get. Money and trading and competition is what makes the world go round. It's always going to be a higher priority than the quality of the night sky for stargazers, or astronomical research.

For casual stargazing, you might think - well, we got used to electricity pylons and windfarms strewn across the countryside, we'll get used to this too. For scientific research into the cosmos, you might think, well, this has made it harder, and more expensive, but we are still doing some. Move on.

You may choose to agree or disagree with that pragmatic assessment. I am generally a "bend with the wind" kind of person. But I think in this case we damn well ought to make a stand. You don't have to be a Luddite or an Anti-Capitalist to feel that we must stand up to megaconstellations. Giving people better Internet, and keeping Capitalism healthy and competitive, is quite possible *without* thousands of low orbit satellites. Why should we accept arbitrary degradation and pollution when it's not even necessary?

ENGAGE AND MITIGATE? Maybe it's unrealistic to think that we can possibly stop the swarm, but what we can do is to start conversations, try to have an influence, carry out technical studies, and suggest ways we can make things not quite as bad as they might have been. This engagement strategy is the firm stance of many of my astronomical colleagues, especially those who have volunteered for the various working groups set up in the last few months. Some people have been working closely with the space industry for years of course, or are just starry-eyed about the excitement and romance of the space age. Others, seeing the looming problems for their own projects or personal science goals, have made a point of reaching out, starting conversations with SpaceX engineers.

So is mitigation enough? We began this discussion in Chapter 5. Various strategies are under discussion - reducing the reflectivity of the satellites, adding shielding, carefully controlling the satellite orientation during the orbit raising period, publishing acccurate orbit data to allow prediction of sightings, weighing the pros and cons of different orbit heights. SpaceX have even launched some related experiments - the so-called DarkSat and VisorSat versions of their satellites. However, as we saw in Chapter 5, the achievements are relatively modest, and for radio astronomy really not much help at all. Of course, they are definitely worth pursuing. Even moderate improvements may be enough to largely remove the problem for casual naked-eye stargazers - at least as far as the SpaceX Starlink satellites are concerned... but that is not the end of the story.

As we saw in Chapter 7, the problem is not so much SpaceX, as who follows on from them, once the crack is opened and the trickle turns into a flood. How do we know who will be amenable to discussion and improvements, and who will ignore our polite email requests? Once the market becomes competitive, corners will be cut and methods will become more ruthless, as companies try to get the crucial margin. That's how Capitalism works.

PROTEST? THEY SAY THAT YOU NEED a long spoon to sup with the devil. Well, maybe no spoon is long enough; sitting down to dinner with rich and powerful forces can often be a mistake. They will offer us some pie, and start bulldozing our homes while we are at their table. Dramatic I know, but some of my colleagues really are very angry and suspicious.

Outside of elections - a blunt tool - public protest is the main way that citizens at large, or concerned sections of the community, try to get bad things changed. Of course, not every country in the world makes either of these things - elections or protest - easy for their citizens, but let's go with this. So you make a fuss, set up a website, start a petition, get on the TV news, maybe organise a march. You hope that your government will notice and care. If making your feelings known is not enough, you might choose to begin more disruptive protest - camp outside the nuclear base, become a hunt saboteur, sit in the road and block the traffic. Such methods have a mixed history of success. Protest and disruption dismantled apartheid and got women the vote; but I don't see much sign of nuclear disarmament happening, and the success or otherwise of Extinction Rebellion is still to be seen. But what is certain is that unless at least you make your feelings known, government, and corporate forces will not care.

For the problem of satellite sky pollution, is such protest appropriate? The satellite problem is not as morally important as animal cruelty in its various forms - factory farming, or fox-hunting, or badger culling. It is not as important for the future of humanity as climate change, or the loss of biodiversity, or the problem of plastic in the sea. But in fact for me it is closely linked to those things. I see the streaks across the sky, and the lurking danger of advertising, as the final damned straw in our relentless destruction of the natural environment by uncontrolled human activity. This applies even more clearly to the the dark-sky problem, and so to sky pollution more generally; but the satellite pollution problem is global; it's an annexation of the sky.

Fig8.1: Online petition protesting the damage caiused by satellite megaconstellations, set up at the Avaaz site by Gaspar B. (See Appendix)

Taking the first step - increasing public awareness and making a fuss - seems absolutely right. Is more strident direct action justified? Everybody should make their own mind up. For me, protest about satellite mega-constellations only makes sense if tied together with broader environmental issues. But the problem for effective protest is the same as for climate change - the problem is not local, or national, but global. How do you convince all those governments at once, who will worry that if they are too hard on business, other countries will exploit this as a weakness and grab an advantage?

LITIGATE? IN MANY COUNTRIES THERE ARE LAWS AIMED at protecting people from harm caused by others. Of course if the injured party is an individual, and the injuring party is a corporation, this is not easy. Litigation is expensive and corporations have deep pockets. In the USA, a tradition grew up of class action lawsuits, where a large group of people can be represented by one party. There have been many famous and succesful examples, such as the case against PG&E for dumping contaminated water in Hinkley California (the "Erin Brockovitch" case); or Roe vs Wade, which established the principle that restricting a woman's right to choose abortion was an invasion of her privacy, covered under the ninth and fourteenth amendments to the U.S. Constitution; or the claim concerning the 2010 Deepwater Horizon Oil Spill, which affected the lives and livelihoods of thousands of people. The idea of such group actions has been gradually spreading to other countries, and as I write, developments are underway to enshrine something similar in EU law.

Could such a claim be taken against SpaceX and others for the damage caused by satellite constellations? It is difficult but not impossible. Who would be the injured parties? The stargazing public? Individual professional astronomers? The funding agencies who paid for the facilities whose measurements have to be expensively repeated? What is the damage caused? The loss of beauty? Some number of hours of telescope time, costed at some agreed rate? The increased risk of the destruction of your spacecraft? Under what law is this harm protected against? The law in which country?

Again, we see that the difficulty, just as with pursuing the protest route, is the global nature of the problem. The cause of the damage may start in one country, but produce its damaging effects somewhere else. Who has the right to pursue a claim against a corporation in a different country? Although the legal right to pursue representative actions is becoming common across much of the world, as far as I am aware there is no tradition of, or mechanisms for, individuals or groups - as opposed to nation states - pursuing trans-national litigation. Anybody fancy the the chances of astronomers in France suing a corporation in China?

Maybe it is worth pursuing legal rights for individuals or groups within single countries - especially the USA, where currently most of the problem resides, and there is the strongest tradition of class action. But for a real solution, we have to think about international law.

REGULATE? IF THE PROBLEM with either protest or litigation is that these methods are local, while the problem is global, then surely what we need is international regulation? Well yes, but how is this to be done? Our world is carved into "sovereign" states. These are the entities that set laws, and have the state apparatus to enforce them. Citizens and corporations of Tanzania obey Tanzanian law, whatever their neighbours in Mozambique may think. When we talk of "international law", what we mean is agreements between nation states to follow common standards and rules. Federations like the European Union and the United States of America seem at first to blur the lines, but the principles remain clear. From the international point of view, the USA is a single sovereign state, whereas the EU is collection of collaborating states that retain their own sovereignty. France remains a sovereign state, but Massachusetts does not.

Some international agreements are essentially just a matter of custom and practice, backed up for example by a network of trade rules and tariffs. Some agreements are encoded in formal treaties. Often, such treaties - like the Geneva Convention - require each signing state to embody the treaty provisions in their own national laws. In other cases, international bodies are created, to which some degree of sovereignty is ceded - for example the International Criminal Court, the World Trade Organisation, or of course the United Nations.

As my colleague Martin Elvis pointed out to me, an interesting exception to the general rule of the world being run by agreements between states is the way that the Internet, and the World Wide Web, are regulated, by a combination of the Internet Engineering Task Force (IETF), the Internet Corporation for Assigned Names and Numbers (ICANN), and the World Wide Web Consortium (W3C). All of these are to varying degrees informal and self-organising, but highly structured, with members who are a mixture of individuals and organisations. Somehow it all works, without governments and laws.

So where do astronomy, the beauty of the night sky, and the dangers of space debris fit into the international regulatory framework? Astronomy has a body that sets its own internal standards (the International Astronomical Union), and there is a body that *lobbies* for the darkness of the sky (the International Dark Sky Association) - but whose job is it to *protect* the sky from others? We need to take this in three steps - how radio astronomy is in principle protected, how the space industry is regulated, and how optical astronomy is protected. The last one is easiest. Lets deal with it first. Who protects and regulates optical astronomy? Nobody.

Now... radio astronomy.

THE INTERNATIONAL TELECOMMUNICATIONS UNION (ITU) is an organisation within the United Nations which sets communications technology standards and allocates bands within the radio frequency spectrum to classes of activity. As well as 193 nation states, it has 900 "sector members" including companies, universities, and other international organisations. The radio astronomy community has been actively involved since its beginnings in the 1940s, and today several radio astronomy organisations and projects are members. This takes a lot of effort - added up over many individuals, it requires roughly the equivalent of 10 full time staff dedicated to dealing with ITU matters. But it has been worth it - the regulation RR5.340 defines prohibited bands to protect radio astronomy, and RR5.149 requires operators to "take all necessary steps" to protect radio astronomy.

This system is good but imperfect - compliance is far from complete, and in practice radio astronomy needs to stray outside its protected bands, as science evolves. For example, radio astronomers now want to study the emission from Hydrogen gas from a time soon after the Big Bang. Because of the expansion of the universe, the classic 21cm Hydrogen signal is redshifted into frequencies that aren't formally protected. This wasn't foreseen decades ago, but now is one of the most exciting problems in our study of the origin and evolution of the cosmos. Two things have kept the imperfect protection system workable. The first is that, until now, most of the satellite interference has come from spacecraft in geosynchronous orbits; these objects have slow moving predictable tracks. The second thing is that many nations designate protected geographical "radio quiet zones", which the ITU community recognise. Radio astronomy takes advantage of this by building telescope arrays in the Australian outback, or the Atacama desert. Both of these safeguards - predictable tracks, and quiet zones - are destroyed by low orbit satellite constellations. They move fast and unpredictably, and are over your sky no matter where you are.

The ITU takes an explicit role in satellite communications. Operators are required to avoid harmful interference with other operators, whether other satellites, or facilities on the ground. I know that my radio astronomy colleagues are gearing up to fight that corner. Of course radio astronomers are not the only stakeholders - many satellite companies, including SpaceX and OneWeb, are also sector members. Furthermore, the ITU also has an explicit "development" remit, to improve global access to communications technologies, and bridge the "digital divide". The astronomy community really should not fight against that remit; rather, we need to argue that placing tens of thousands of satellites in low earth orbit is *not necessary* to meet that remit; and indeed will cause pointless damage to other operators.

IN PRINCIPLE, BEHAVIOUR IN SPACE IS REGULATED by the Outer Space Treaty of 1967. Above our heads is the "air space" that comes under the jurisdiction of each sovereign state, but from 100km up is "Outer Space". (The exact height that Outer Space begins is not legally defined; 100km is the usual convention.) The Treaty declares that outer space should be used "for the benefit and in the interests of all countries, irrespective of their degree of economic or scientific development, and shall be the province of all mankind". The Treaty is quite short, as are the four more follow on treaties that have been debated and (partially) agreed at the United Nations over the years. The Treaties set out very general principles. Space is free for all to explore and exploit, so that no company or nation can claim ownership or plant a flag. Military activities are restricted - roughly speaking you can spy but you can't shoot. Bodies launched into space remain the property of the launching states or organisations, so you can't salvage somebody else's defunct satellite. Everybody should keep a register of what they have launched into space. Each nation is - in principle- liable for damage caused to others by its own state agencies and private companies.

This is all pretty loose. It is expected that states embody the principles of the Outer Space Treaties in their own domestic space law, and regulate the activities of the agencies and private companies within their jurisdiction in the spirit of those Treaties. Likewise, the space industry community is effectively assumed and expected to develop its own industry associations and develop and follow appropriate codes of practice. However, only a minority of states, about 25, have any space law at all, and their codes vary enormously. Unlike the ITU, with its volumes and volumes of detailed specific regulations that the representatives of 193 states and 900 organisations have thrashed through and are required to follow, the Outer Space Treaty provides only very broad guidance through seventeen one paragraph articles. You can read them all in a quarter of an hour.

What do you do if think you have suffered some damage from the activities of a space company? The Liability Convention tells us that if you live in the same country as that company, you should pursue a claim through your own domestic laws. If you live in a different country, you have to ask your own state to pursue a claim through diplomatic channels on your behalf. To my knowledge, that has successfully happened only once. In 1978 the Soviet Union paid compensation to Canada, following the fall of the Soviet nuclear reactor Kosmos-954 onto Canadian soil. What about the growing problem of space debris? There is at least a growing sense that the problem needs to be taken seriously. The key forum for discussing the problem is the International Space Debris Co-ordination Committee (IADC). But there are no hard legal consequences for bad actors.

This loose and co-operative approach, and the lofty principle of space being for the benefit of all humanity, made sense in 1967. But "Free For All" can fast turn into "Anarchy in The Wild West". Three key things have changed. First, there is big money to be made. As well as selling low latency connection to banks, and satellite dishes to rural gamers, there are valuable resources waiting to be extracted from the Moon and the asteroids. Second, the crowding of space is bringing the problem of debris to a tipping point, threatening to make space activity unsustainable. Third, activities in space are beginning to cause what should be seen as general environmental damage, as well as specific damage to scientific activity.

As the financial stakes in space activity increase, things are bound to heat up. The telecommunications company Viasat recently submitted a filing to the US Federal Communications Commission which was effectively a complaint about the activities of SpaceX. This is just the first signs of what could soon become serious capitalist warfare, crossing international boundaries. Who could rule in such disputes as they become more serious? There is no International Space Court.

SO CAN ANYBODY DO WHAT THEY LIKE IN SPACE? Certainly not. Most people working in Space Industry might even complain that it is tediously over-regulated. First and foremost, agencies and companies have to follow the licensing and regulatory procedures in their own countries. Secondly, you have to comply with international communications regulations. In practice this means working with your own national telecomms agency, and also with the ITU, in a complicated loop.

For satellites in those high-up Geostationary Orbits (GSO), there are only a limited number of orbit slots as well as frequency slots that can be used without interfering with other satellites. Roughly speaking, each nation negotiates for a number of slots which it can then hand out. (This has led to the problem of "paper satellites"... but thats another story). This control of GSO slots has led to a kind of assumption that the ITU "owns" space regulation, and that it is appropriate for the ITU to develop policy on low-orbit constellations as well. Certainly OneWeb or SpaceX or Telesat can't launch without ITU approval. The ITU are clearly aware of the debris problem, and the distress that Starlink satellites have caused optical astronomers, but its not obvious that expanding the remit of the ITU is the right way to address these problems.

IN MANY WAYS, THE SEAS OF THE WORLD are much like outer space - a shared realm surrounding our nation states that we launch onto, travel around, and exploit for commerce, exploration, and scientific resesarch. Historically, the accepted approach was the doctrine of the "freedom of the seas". A three-mile strip of sea around your coastline was under the jurisdiction of your own state, but the open seas were deemed to be free to all and belonging to none, so that all could explore and exploit without interference by other states. Sound familiar?

During the twentieth century this system began to break down. Fishing vessels could stay on the ocean for long periods of time and travelled far, fishing in regions that other communities had informally assumed were effectively theirs. As stocks dwindled, tensions rose. More minerals and oil were found on the seabed, increasing the value of the sea. Some nations claimed coastal zones of twelve miles, or even two hundred miles. The USA asserted rights to natural resources over the whole of its continental shelf. By 1967, just as the ink was drying on the Outer Space Treaty, the maritime community had decided that simple "freedom of the seas" was not fit for purpose in the modern world, and the United Nations set about to fix it. The Convention on the Law of Sea established a comprehensive and detailed set of laws and regulations governing all uses of the oceans and their resources. Even more important, the Convention itself specifies dispute resolution mechanisms, and the UN established the International Tribunal for the Law of the Sea, as well as the International Seabed Authority. Those organisations have teeth.

So it seems that space regulation may be going through the same historical process. We are rapidly arriving at a time where we need to protect the space environment, recognise the wider impact of space activities, and establish a mechanism to resolve disputes. Another interesting precedent is the situation of Antarctica. The Outer Space Treaty of 1967 was closely modelled on the Antarctic Treaty of 1961, and its careful and principled establishment of Antarctica as a resource for all to share. There are however two crucial differences. First, the Antarctic Treaty explicitly gives all nations the right to monitor what other nations are doing. Secondly, Article XI of the Antarctic Treaty specifes a dispute resolution mechanism. This is not as grand and organised as the International Tribunal for the Law of the Sea, but at least there is something. It states that first the disputing parties should try to resolve the matter by negotiation, arbitration, or whatever means necessary; but if that fails, they should bump the matter up to the International Court of Justice.

SO WHERE DOES ALL THIS LEAVE US? Looking at the options we have examined, my answer would be "all of the above" - except the first one, roll over and give up. Ultimately, I do feel that international space regulation needs a serious shake-up, along the lines of the law of the sea, or all of astronomy, space science, space industry, and space exploration are heading for serious trouble. However... it took nine years to debate and conclude the Convention on the Law of the Sea. Meanwhile, we absolutely should publicly protest, engage and negotiate, and possibly litigate. These things, in my mind are not inconsistent, but all part of the pressure towards evolving a system that works for all.

IX. A CHRISTMAS TALE

Jan 7th 2021

THE YEAR HAS TURNED but the pandemic continues. Scotland, like much of the world, is back in lockdown. Christmas was quiet, and New Year's Eve - the Scottish Hogmanay - was eerily restrained. No massive street festival. No youthful parties and drunken citizens. No neighbourhood first footing with coal, cake, and whisky. We managed to briefly step outside and raise a glass to carefully distanced neighbours, but the street was almost deserted.

It feels particularly strange because the Christmas-Hogmanay period in Edinburgh has grown increasingly controversial for its massive scale and commercialism. Some years back the City Council starting sealing off the centre of town at Hogmanay. They were probably wise to do this, as the crowds were becoming dangerous. More controversially, they took the opportunity to milk it. You had to buy a ticket. But if you did there were concerts and fireworks galore. Likewise, what started as a small "German Market" in the run up to Christmas grew and grew over the years. In late October 2019 I went shopping in Princes Street and tried to walk along the beautiful East Princes Street Gardens. But it was fenced off. The company contracted to run "Edinburgh's Christmas Market" - Underbelly - were making their preparations.

Dozens of trees had been cut down and acres of temporary structures were underway. By January the structures were gone and what was left behind was a muddy wasteland. The Council were busy returfing and seeding but it was early Spring before the gardens were nice again.

Fig. 9.1 Oct 2019: the way is barred to East Princes Street Gardens. Strange, I thought this belonged to me, as a citizen of Edinburgh...

The experience of being blocked from the gardens in October really threw me, as it did many other citizens. "What?" we thought. "Can they do that? These gardens are public space - they belong to all of us. They can't just close them down for a quarter of the year". Much complaint burst out on social media and local TV. I certainly wrote an angry tweet or two. But I felt awkward doing so. Like many thousands of others, I have enjoyed the Christmas Market over the years - shopping for gifts, skating, taking the kids out, showing visitors around, and just sniffing the gluhwein in the cold. At first much of the debate on social media painted the situation as being one of an argument between elite puritan spoilsports on the one hand, and ordinary people having fun on the other. But soon it became clear that in fact large numbers of citizens, like myself, were on both sides at once. People were starting to say "well, ok, but - there must be a better way to do this? Surely we can have Christmas fun, and allow the Council to make some money, without trashing our gardens?"

BECAUSE OF SARS-COV-2, there was of course no Christmas Market in the year 2020. So solving that social puzzle has been put off. Likewise, there was no ticketed Hogmanay Street party. But Underbelly found a novel way to make a kind of New Year civic celebration. Over several successive nights they launched 150 luminous drones into the night sky in various locations, and controlled their motion to make amusing moving displays - smiley faces, prancing reindeer, hands of friendship joining across the sky, etc. Not many people actually managed to catch the displays I think, but a thirteen minute video was put together, complete with words by Edinburgh poet Jackie Kay, read by famous actors. Most people were intrigued and impressed, and indeed it was very nicely done. But some of us were at the same time worried, because of the precedent that had been set. How long before the Council realises they could make some money by licensing drones that advertise a fizzy drink? Maybe a different fizzy drink every week? How long before the richest fizzy drink companies press to be allowed to launch their own naked eye satellite constellations? Yes, it would cost a lot, but they could reach billions of people at once.

THE ANALOGY BETWEEN our local Christmas market problem and the global problem of environmental pollution by satellites is hopefully pretty obvious by now. Once again I find myself on both sides at once. As I have strained to explain throughout this book, I love space, I love technology, and I benefit hugely from the Internet. Why should I complain about plans to connect up the world? But just as finding myself blocked from the gardens was an unexpected shock, so was the sight of bright streaks across astronomical images. Once again I found myself thinking "Wait, what? Are they allowed to do that?" and then also "surely there is a better way to do this without trashing the garden?"

On the first question - can they do that - the answer is quite worrying. Although there is some international co-ordination, satellite launches are licensed by national governments, and seen as either a telecomms issue, or an economic growth issue, or both. In other words, if you are, for example, a US-based company, the people you have to convince are the US Federal Aviation Authority (FAA), and the US Federal Communications Commission (FCC). However, far from trying to hold projects back while we look at the consequences, the FCC just awarded almost a billion dollars to SpaceX to subsidise Starlink. It will be a tough fight to get the brake pulled on that train.

On the second question - is there a better way - the answer is surely yes, but it may be hard to get there from here. Recently I watched a series of online talks hosted by the ITU, from companies providing or planning to provide satellite Internet. All of them spoke passionately about global connectivity - surely in the year of the pandemic, they argued, we can see more than ever how important connectivity is, and how unfair it is that much of the world is not connected? Well, who'd argue with *that?* But it really doesn't follow that to achieve these noble aims we need tens of thousands of satellites a few hundred kilometres up. Or that we really need 50msec latency, as opposed to 150msec. Or that we need multiple competing systems. Yes, I know that's how free enterprise is supposed to work - open up an area for competition, and let the market sort out the weak from the strong. But the world is not in truth a seamless free market. Each country will want to push their own companies. But if it really is a global fairness issue, and space is for the benefit for all the people of the world, as opposed to being for the economic benefit of the few nation states with launch capability, why can we not slow down, debate, and commission a truly global shared infrastructure?

ONE FINAL WORD. I am repeating myself, but it bears repetition. I am not foolish or selfish enough to think that the damaging of astronomical science, and the pollution of the night sky for all, is as important as climate change, the loss of biodiversity, or the cruelty inflicted on many thousands of animals every day. But it is important; we should care; and I see it as emblematic, and symptomatic, of the environmental damage caused by thoughtless human activity and greed, and with much the same root causes as those other big issues.

The sky matters. Lets not lose it.

APPENDIX

I HAVE TRIED TO KEEP THE TEXT in this book straightforward and understandable, while hopefully still accurate, but you will I am sure understand that this hides a lot of technical detail. Some things really are simple in essence, and I feel that a blunt simple description is actually in some ways better. In other cases the simple approach papers over quite a few bumps, wrinkles and cracks. If you want to really understand, you might want to follow up with some more detailed reading. Below I provide some links to the web pages of key organisations; some general further background reading; and for the brave, links to books and to technical articles - scientific papers by my colleagues in research journals, and reports of technical working groups. These really deliver the goods, but tend of course to be written in dry, cautious, and sometimes almost impenetrable language.

As well as (hopefully) providing an understandable description and analysis of the issues, I have tried to provide a broad overview of the situation. I enjoy doing this, but at the same time feel a little guilty, because I am shamelessly presenting clever things done by other people, especially many of my colleagues who have worked hard on a variety of working groups, and continue to do so. There isn't room to name them all one by one, but below I provide links to reports already made by some of these working groups.

The external links to the various websites and articles are listed in the "Bibliography" section, where they are spelled out in full, so that the links are useful in the printed version of the book as well as in the ebook version. However, if you are reading the ebook version, you can click on the internal cross-links given in this Appendix to jump to the listed external link in the Bibliography.

Space Industry

WHO ARE THE KEY ACTORS in this issue of satellite constellations?

The main SpaceX website is at [URL1], but note that there is a separate website for their Starlink programme [URL2]. In a similar fashion, their UK-based rivals OneWeb have a general website [URL3], and another website specifically for their OneWeb satellite programme [URL4]. Other companies/agencies known to be planning satellite constellations include the Canadian Telesat company [URL5], the Chinese Aerospace and Industry Corp [URL6], the Russian agency Roscosmos [URL7], and yet another US company called Globalstar [URL8]. Meanwhile, this blog post by Larry Press [URL9] states that three other Chinese companies are planning low-orbit fleets, and this news item in TASS [URL10] states that a Russian company called "Success Rockets" plans 500 satelllites. Success Rockets are too new to have a website yet. Meanwhile, this news item in Arstechnica [URL63] states that Russia is planning a Starlink rival called "Sphere" and even planning to fine citizens who use Starlink services! Next up, Amazon's plans for 3,236 satellites are being routed through a wholly owned subsidiary which does not yet have a website - Kuiper LLC. (See this news item [URL11] and this wikipedia page [URL12]). The US company Lynk [URL13] are planning global cell-phone connectivity via a constellation of low earth orbit satellites. Meanwhile, after a bad start described in this news article [URL14], Facebook is now operating an experimental satellite [URL15]. It isn't yet clear what Facebook's overall ambitions are. What might this all add up to? Here is a Youtube video [URL16] made by the folks from Analytical Graphics Inc [URL17], describing their compilation of sources indicating that 107,000 satellites are planned to be launched by 2029. Last but not least, here is a webinar [URL18] hosted by the ITU where some of the key players present their plans - SpaceX, Amazon, Telesat, OneWeb, and SES. It is very interesting but over two hours long!

It is worth contrasting these plans for thousands of low-orbit satellites with the fact that there are already companies offering internet service to remote locations.Viasat [URL19] offers high-speed (but not low latency) internet service to rural locations in the USA using a single geostationary satellite. O3B [URL20] offers global internet connectivity, with around 20 satellites in Medium Earth Orbit, about 8000 km up. The parent company is SES networks [URL21]. The development of this programme is described in this wikpedia page [URL22]. Given that these projects already work, the only reasons I can see to develop constellations of thousands of low orbit satellites are (a) because your business model needs the absolutely shortest possible latency, or (b) because you have a competitive urge to capture market share.

As I mentioned several times during the book, SpaceX are at least engaging with the scientific community, and OneWeb have an initiative they call "Responsible Space"

[URL23]. This seems nice, but of course ignores the fact that the obviously responsible thing to do is not to launch the satellites at all... There is also an interesting initiative called Gnosis [URL24] which brings together scientists and industry to "understand and solve problems related to the ever growing problem of space debris". These initiatives are I think a good start, but a bare start.

There are a number of space industry associations that help to foster the interests of satellite companies, and are also important in the legal and governance issues we discuss below. These include the Satellite Industry Association [URL25], the Space Data Association [URL26], the Satcoms Innovation Group [URL27], the European Satellite Operators Association [URL28], and the Global VSAT Forum [URL29].

Space Object Tracking

HOW DO WE KNOW WHATS UP THERE? First, launches and re-entries. There is no internationally standardised method of registering launches. Mostly, you just have to rely on press-releases and other public announcements by the various agencies and companies. However the the US and Russian militaries do have a system of notices called ILAMs (Initial Launch Alert Message) which basically say "if you spot this, don't worry, it's not an attack". There are also NOTAMs (Notice To Airmen And Mariners) which say "you might want to stay out of this area of sea for a bit", in other words providing warning of rocket stages on the way back down.

Things are a little better for the tracking of things which are already up there. Many agencies and companies publicly distribute orbital data for things they have launched. Most people these days do this by sending the data to Celestrak

[URL30] which is a system set up and run by T.S.Kelso. Third parties then access the orbital data from the Celestrak database. The orbital data is in a very old fashioned format called a TLE (Two Line Element), which is two lines of ASCII text, each 80 characters long. Old timers will understand why there is an 80 character limit... I leave this an exercise for the reader. Anyway, this is enough to pin down the location of a satellite, or piece of debris, to about 1km. For objects in high orbit, the orbital data stays accurate for a long time. For an object in low orbit however, the information needs regular updating, as even the tiny amount of atmospheric drag there is at a height of 500km is enough to gradually change the orbit.

For ongoing tracking of everything up there, the key source of information is the 18th Space Control Squadron [URL31] of the US Space Force. They operate the US Space Surveillance Network [URL32], a system of radars and telescopes which track everything that moves. Out of about 20,000 objects tracked, they release orbital data on all but about 500, which are military secrets of course. The orbital data is made available through a commercial contractor called Space-track [URL33]. You have to register to get access so they know who you are, but it's otherwise pretty much open. What they release are TLEs, just like at Celestrak, in the form of the "US Satellite Catalogue". The data are good enough for most purposes, but of course not good enough for an Anti-Satellite missile to hit the object concerned. Of course internally, the US Air Force has much more accurate orbital data. From this more accurate data, the Air Force calculates whether two objects are likely to collide, and sends alerts to the owners, through Space-track. This is getting harder every day, as the sky gets more crowded, and the space industry more various.

The near-miss of Starlink-44 and ESA's Aeolus satellite (see [URL80]) apparently happened because Space-X "lost the email".

Everybody is and should be grateful to the US surveillance programme, not least because matching up individual detections to persistent objects is a highly non-trivial technical problem. The Russians have a similar programme, which seems to be partly public [URL34]. It is not clear whether or not the Chinese have an equivalent programme. There is an equivalent European research programme, but it has not been operationalised. Meanwhile, commercial operators are beginning to set up their own surveillance systems - Leolabs [URL35] for low orbit objects, and ExoAnalytic [URL36] for objects in GEO orbits. Finally there are a number of hobbyist satellite watchers. There is a nice wikipedia page [URL37] about them. The secret satellites not released by Space-track are often really bright and easy to spot, so those hobyists can tell you all about them.

Along with Celestrak, the other intermediate source that a lot of people rely on is Jonathan McDowell, who has been the key technical consultant for this book. In the guise of "Planet 4589" McDowell maintains a web page [URL38] that is an extensive resource on the history of space exploration, issues a monthly Space Report [URL39], and tweets several times a day keeping us up to date on the latest launches and other events. There are two key things about McDowell's activity. The first is that he maintains a complete history - he has information on all past launches and objects, not just current ones. The second is that, rather than relying purely on the US Satellite Catalog, he aggregates information from many different sources, and interprets and explains what he is seeing. In 2020 he published the General Catalog of Artificial

Space Objects (GCAT: [URL40]) that makes public the database underlying his Space Report. His Internet persona refers to the Minor Planet named after him by the Minor Planet Center, in gratitude for his work. Asteroids are not his normal line of work (or mine!) but it is interesting to note that they are typically discovered by seeing unexpected fast moving objects in the sky, or equivalently, streaks in integrated images...

Finally, there are a number of third party websites which use all this public information to provide interactive viewing of satellites in real time - for example, Satview [URL41], and N2YO [URL42]. My favourite is In the Sky [URL43], which is a much more general night sky guide, astro news and interactive planetarium-style facility, as well as providing satellite viewing.

Space Information and News

THERE ARE LOTS OF OPTIONS for keeping up with general news about space, satellites, and launches, so I will just list a few. First, you can take a look at the news sections of the web pages of the major space agencies, such as the European Space Agency (ESA: [URL44]), the US National Aeronautics and Space Administration (NASA: [URL45]), the Japanese Aerospace Exploration Agency (JAXA: [URL46]), or the Russian Space Agency Roscosmos [URL47]. Then there is a variety of magazine style websites - Space News [URL48], Spaceflight Now [URL49], Space.com [URL50], Spacewatch [URL51], and Skyweek Zwei Punkt Null [URL52]. For particularly serious and well informed space news, and more general information and analysis, there are two excellent resources: one is Jonathan's Space Report [URL39] and the other is Space Policy Online [URL53].

Of more general web news outlets, some have particularly good space news coverage. Examples include this article from Business Insider [URL54] about the failure rate of Space X launches, and this article from Advanced Television [URL15] about Facebook's entry into space.

Most of these news websites have associated Twitter accounts of course, which makes it easier to keep up to date, at the usual risk of distraction and information overload, so I will restrict myself to mentioning two Twitter accounts: one is of course @planet4589 [URL55] and the other is @markmccaughrean [URL56], Senior Science Advisor at the European Space Agency. In the manner of the best Twitter accounts, as well as getting the latest ESA PR from Mark, you will get a very personal spin on all sorts of issues.

Space Debris

FOR A GENERAL OVERVIEW OF THE PROBLEM OF SPACE DEBRIS, a good starting point is the Wikipedia entry [URL57]. If you want more detail, both ESA and NASA run orbital debris programmes. The public home page for the ESA Debris Office can be found at [URL58]. You will find lots of information, images, and news stories, and there is also a playlist of videos at [URL59]. The "Space Debris 2013" video is a good introduction, and the "Journey to Earth" video is a particularly useful perspective. To get more detail you might want to look at the page on the Current State of Space Debris [URL60] where you will also find a link to ESA's full technical Annual Report on the Space Environmemt [R1]. A lot more technical detail and tools can be found in the debris pages of the European Centre for Space Law [URL61], and the ESA

Space Debris User Portal [URL62].

The NASA Orbital Debris Program Office (ODPO) is part of the Astromaterials Research and Exploration Science program (ARES), and has its home page at [URL64]. The Office of Safety and Mission Assurance (OSMA) also takes a role in Orbital Debris, and has a web page here [URL65]. The ODPO runs programs in measurement, modelling, protection and mitigation, and issues a quarterly newsletter [URL66]. It also defines standard practices in debris mitigation which US government offices are obliged to follow [URL67]. Their web pages also have some nice graphics and videos, including one in the Modelling page on the growth of debris over time [URL68], and another on the Protection page with a very nice 3D fly-through of a model of the debris population [URL69]. A useful summary of the work of the ODPO is in this wikipedia entry [URL70].

Following the worries raised by Don Kessler, whose personal website is here [URL71], NASA were the first agency to develop mitigation guidelines, but other national and international agencies soon followed suit. The UN Committee for the Peaceful Uses of Outer Space (COPUOS) maintains a list of national guidelines [URL72]. Approaches to the debris problem are debated at the Inter-Agency Debris Co-ordination Committee (IADC) which holds regular meetings [URL73]. This process eventually produced an agreed set of guidelines endorsed by COPUOS [URL74]. Note however that these are guidelines, not regulations.

For more on the Kessler syndrome, there is a fascinating article by Raffi Khatchadourian in the New Yorker magazine [URL75], and there is an excellent Wikipedia entry [URL76]. There is a very nice article on debris tracking by Agnieszka Gautier on the EarthData website [URL77]. There are more interesting articles on collisions, debris, and the space industry at [URL78],[URL79], and [URL80].

On the more technical side, you might want to look at a research paper by Honquiang *et al* 2020 [R2] which, starting from the recent near-collision between Starlink-44 and ESA's Aeolus satellite, assesses the likelihood of future collisions, or another paper by Blake *et al* 2021 [R3] which discusses a ground-based survey of debris in Geosynchronous orbit.

Research papers in scientific journals

NOTE THAT MOST OF THE LINKS PROVIDED in this section are to the NASA Astrophysics Data System (ADS) service, rather than to the journal in which the papers are published. This is now the standard way that astronomers link to, and search for, scientific papers.

Three recent papers have carefully analysed the general impact of Starlink satellites on optical astronomy; one by McDowell 2020 [R4], and a pair of papers by Gallozzi et al 2020a,b [R5,R6]. The latter pair of papers are explicitly connected with an international online petition called "Safeguarding the Sky" [URL81]. I am signature number 1031 out of 2025 so far.

There is also a short readable article by Tyson and Parriott 2020 in *Science* [R7]. A paper by Hainaut and Williams 2020 [R8] specifically considers the impact on Europe's premier astronomical facility, the European Southern Observatory (ESO), and a paper by Tyson et al 2020 [R9] focuses on the Rubin Observatory, currently under construction - a wide field survey telescope which the international community has been keenly looking forward to, and which will suffer particularly badly. Two papers by

Tregloan-Reed et al 2020a,b [R10,R11] report brightness measurements of the Darksat experimental satellite launched by SpaceX, and a paper by Mallama 2020 [R12] reports the brightness of OneWeb satellites.

The paper by Rawls et al 2020 [R13] is written by astronomers, but is largely a sceptical - and cogent - analysis of the claimed business case for Starlink that it will connect up poor and underserved parts of the world.

Study Groups and Reports

A NUMBER OF SMALL OR MEDIUM SCALE technical studies have been going on, which have led to the scientific papers listed above. The papers by Gallozzi *et al*, by McDowell, and by Hainaut and Williams are especially thorough and comprehensive. However, soon after the initial alarm felt by astronomers as they saw the streaks appearing in their images, several major efforts got underway, pooling the work of astronomers worldwide.

The unifying body for such work is the International Astronomical Union (IAU). This is the organisation which defines and agrees international standards in astronomy - how we express co-ordinates, what the definition of a planet is, and so on, in a series of (often rather tedious!) commissions, committees and working groups. But it also has a remit to "safeguard the science of astronomy in all its aspects". The IAU quickly set up a Satellite Constellations Working Group (SCWG) [URL82]. Their web page has a lot of useful information, and also links to many other activities.

The American Astronomical Society (AAS) also got organised quickly, and liased with the international SCWG. They maintain a web page with live updates [URL83] on the satellite constellation situation. This has lots of very useful and reliable information. In June 2020 they organised a conference with international attendance. This led to a comprehensive technical report known as the SATCON-1 Report [R14], which includes recommendations as well an analysis and possible mitigations. This is currently the definitive technical statement of the situation. A number of other national working groups are underway. In my own country for example, the UK, a working group has been set up by the Royal Astronomical Society [URL84].

The other significant activity we need to mention is a study being led by the IAU, Spain, and the US NOIRLab, leading to a report to the UN Committee on the Peaceful Uses of Outer Space (COPUOS - see more about this in the "legal framework" section below). This study is broader than the satellite megaconstellation problem, addressing also the dark sky problem, radio interference in general, and the effect of light pollution on the bio-environment. This study was meant to be built around a large international conference in the Canaries called "Dark and Quiet Skies". Because of the pandemic, the conference has been delayed to October 2021. However, a preliminary online version was held in October 2020, which was attended by 970 astronomers, including myself. The conference website is here [URL85], and a very thorough report has recently been completed [R15]. Probably even more important, a digest of the report has been constructed as a "Conference Room Paper", submitted to the April 2021 meeting of COPUOS [R16].

Also worth mentioning in this section is that Jonathan McDowell has recently issued the 2020 version of his annual report of Space Activities [R17]. For completeness here it is also worth listing the ESA Annual Space Environment Report

[R1] which we already mentioned in the Debris section.

Dark Skies

THIS BOOK IS ABOUT THE POLLUTION CAUSED by satellite mega-constellations, but we have referred often to the issue of light pollution more generally, and the problem of preserving dark skies. Here I will provide just a few links to where you can find more about the efforts to protect dark skies. The most up to date and definitive technical statement is in the "Dark and Quiet Skies" report [[R15], which we have already referred to above. The key link for all matters to do with dark sky protection is the IDA - the International Dark-Sky Association [URL86]. The IDA does all sorts of policy work, education and outreach, makes recommendations on sky-friendly lighting, and also formally recognises "Dark Sky Parks" all over the world. The fourth such Dark Sky Park, and the first in the UK, is not that far from me - the very beautiful Galloway Forest Dark Sky Park [URL87]. There is also a dedicated Dark Sky Scotland website [URL88], and I am sure there will be something similar near you. You could also try contributing to Globe at Night [URL89], which is a Citizen Science project where people measure sky brightness in their locality and upload their measurements.

If you want to dig deeper into the Dark Sky issue, I recommend this wonderful online presentation by Diana Umpierre [URL90]. This presentation makes it really clear that it's not just about astronomy; light pollution is about safety, wildlife, health, and energy waste as well. There are a number of research papers on sky pollution, such as this paper by Bustamante-Calabria *et al* 2021 [R18] on how the pandemic has affected urban light emissions, or this one by Gallaway, Olsen and Mitchell 2010 [R19] on the economics of light pollution, or this fascinating recent paper by Kyba *et al* 2020 [R20] reporting an experiment conducted with the co-operation of the City of Tucson, who temporarily reduced their streetlighting, while a satellite measured the difference from space. The result found was that streetlighting is only 13% of the problem - most light pollution is from advertisements, floodlights, parking lots, and various other sources. Meanwhile, in 2019, the University of Utah launched the Journal of Dark Sky Studies [URL91]. It doesn't have much in it so far, but hopefully that will change.

For keeping up with Dark Sky gossip and news, I recommend following John Barentine's Twitter feed [URL92]. John is the Policy Director of IDA. That sounds dry, but his Twitter feed is far from dry!

Finally, from a UK perspective, the best piece of news for a while is that a cross-party group of Members of Parliament has been formed [URL93] to consider policy on light pollution and make recommendations to the UK government. They have issued their first formal report, which you can find here at [R21], with ten very sensible recommendations, including appointing a designated "Minister for Dark Skies". Fingers are being crossed.

Space Governance

FOR A GENERAL GUIDE TO THE CONCEPTS OF INTERNATIONAL LAW, and the history of class action law suits, see the relevant wikipedia entries [URL94,URL95]. Here is a German news article [URL96] about the new EU laws underway to enshrine the right to collective legal action.

International Space Law is debated and agreed by the COPUOS, the Committee on Peaceful Uses of Outer Space

[URL97], which is administered by UNOOSA, the United Nations Office of Outer Space Affairs [URL98]. UNOOSA has a handy FAQ page [URL99], but there is also a very nice summary and explanation of Space Law and how it works on the website of Space Policy Online [URL100]. You can find the Outer Space Treaty itself, along with the other treaties, as well as the COPUOS guidelines on debris mitigation, and various related General Assembly resolutions, collected together in "International Space Law: United Nation Instruments" [R22]. Also of importance is the International Institute of Space Law [URL101]. There is also an easy to read book, "The Little Book of Space Law" [R23] by Mathew Kleiman, from which I learnt a lot.

As well as the issues of sky pollution and debris, the other reason why space law is likely to heat up over the coming years is the prospect of prospecting - the mining of the Moon and asteroids for valuable resources. Two interesting research papers on this issue are by Krolikowski and Elvis 2019 [R24], and by Elvis, Krolikowski and Milligan, 2020 [R25]. Martin Elvis also has a popular-level book on the topic which is due out soon [R26].

As described in Chapter 8, the only issue over which the Liability Convention has been deployed so far was the argument between Canada and the Soviet Union over Kosmos-954, the Soviet nuclear-powered spacecraft which re-entered the atmosphere and dumped radio-active remains in Nothern Canada in 1978. The final agreement is recorded at the UNOOSA website [URL102], and you can read a fascinating account of the incident, and the "Operation Morning Light" cleanup project, at [URL103], as part of the collected "Operational Histories" of the Canadian Arctic. The USSR eventually paid up three million Canadian dollars.

Talking of polar regions... the 1967 Outer Space Treaty was closely modelled on the Antarctic Treaty, developed between 1957 and 1961. However, a key difference is that the Antarctic Treaty explicitly specifies dispute resolution mechanisms, which the OST does not (see the wikpedia entry [URL104]). The Law of the Sea is in some ways an even better analogy to the regulation of space. Analogous to UNOOSA, there is a UN Division of Ocean Affairs and Law of the Sea [URL105]. Within their web pages, there is a very nice history of how the Law of Sea changed dramatically as commercial and international tensions grew [URL106]. As well as debating and agreeing a new international treaty, over the period 1973 to 1982, two international bodies with real teeth were created - the International Seabed Authority [URL107], and the International Tribunal for the Law of the Sea [URL108]. But are these very formal and superstrict agencies, with their rules painfully thrashed out in arguments between governments, necessary for the smooth and fair operation of an international system? Maybe not. The counter example is the way the Internet works, and how the World Wide Web sits on top of the Internet, through a combination of the IETF, the Internet Engineering Task Force [URL109], ICANN, the Internet Corporation for Assigned Names and Numbers [URL110], and the W3C, the World Wide Web Consortium [URL111]. All these bodies have very clearly defined processes, but they are not run by sovereign states.

So how do you get permission to launch or operate a spacecraft? This is seen as a national matter, but with each nation expected to adhere to the framework set by the various Outer Space Treaties. This means that the methods and the regulations are subtly different in every country. So for example, in my sovereign own state, the United Kingdom, space activities are regulated by the Outer Space Act of 1986, as explained at this government web page [URL112]. The process of getting a license is administered by the UK Space Agency [URL113]. However you also need to make sure you have an allocated radio spectrum slot approved by the International Telecommunication Union [ITU: URL114], which you get by first making a filing to the UK Communications regulatory body, OFCOM [URL115]. In the United States, NASA and the military do their own thing, but commercial space activities are licensed by the Office of Commercial Space Transportation [URL116], a branch of the Federal Aviation Authority. However, once again you need your ITU slot, which you get through making a filing to the Federal Communications Commission [FCC: URL117]. The ITU radio communications sector, ITU-R, has a web page here [URL118], and they have a simple video explaining their work [URL119].

If you take a look at the ITU website [URL114], you can see that it perceives itself as a key international body promoting space as well as regulating it. As I look today, it headlines an article about how space science can further the Sustainable Development Goals of the United Nations. Indeed there is a whole sector, ITU-D [URL120] which is about "ensuring inclusive, equal access and use of ICTs for all" - see also the explanatory video here [URL121]. On October 7th 2020, the ITU hosted a webinar [URL18] featuring presentations by many of the key megaconstellation companies. The tone is very much of warm encouragement rather than being stern gatekeepers. Meanwhile, the USS FCC is actively subsidising internet communications activities, including awarding nearly $900 million to SpaceX for Starlink, as reported here [URL122]. Behind all the warm fuzzy words, the commercial tension is of course building up. This article

[URL123] reports a complaint made by Viasat to the FCC about changes to SpaceX plans.

If space governance is your thing, the Swiss-based SpaceWatch website has some fascinating webinars on these matters. Here is one [URL124] which says that space regulation is too onerous and needs loosening up, and here is another [URL125] which says that we need more laws.

Various other bits of viewing and reading

ONE OF THE EARLIEST ARTICLES in mainstream media about the the threat of megaconstellation was by Martin Rees, the UK Astronomer Royal [URL126]. It is still a cogent short summary of the problem.

If you are interested in the movie referred to in Chapter 1, "A Residence Above the Clouds", featuring Scottish and Spanish students trying to re-create Piazzi Smyth's 1856 experiments, which began mountain-top astronomy, it is available via Chip Taylor Communications here [URL114]. That link includes a trailer.

Here is an extraordinary video of a double train of Starlink satellites immediately after launch, filmed by Olivier Staiger from Switzerland [URL127]. It all seems very beautiful until you realise how these things can photobomb your planned exposure. Here is a link to information about the DES image shown in Fig 5.2 [[URL128]; here is the original of the ruined Hubble Space Telescope image shown in Fig. 5.3 [URL129]; and here is the Guardian report on the Astrophotographer of the Year competition [URL130], featuring the Rafael Schmall image I have used on the cover of this book. Amateur astronomers have set up a website called Satellite Streak Watcher [URL131] to capture examples of satellite streaks, and monitor the growth of the problem over time.

There are some impressive simulations of what Low Earth Orbit and/or the night sky will look like soon. This nice video by Michael Vlasov simulates the naked eye view of 12,000 Starlink satellites [URL132]. This very impressive 360-video by Scott Manley [URL133] shows what the night sky would look like of our eyes could see all the satellites currently orbiting the Earth. Don't forget to pan around! On the professional side, this video made by Rubin Observatory project staff [URL134] shows a simulation of the expected effect on survey images by all current satellites; but by the time the survey starts in 2023, the situation is likely to be much worse.

Seen from space, as opposed to the night sky view from the ground, here is a simulation by the folks at AGI of all 107,00 satellites for which FCC and/or ITU filings have been made which might be operational by 2029 [URL16]. AGI have run a series of marvelous videos about space related issues, which they call Spacecasts. Here is Spacecast 27 [URL135], which features T.S.Kelso showing how the Celestrak data can be used to show the tracks of Starlink spacecraft.

For an explanation of how mega-constellations could impact astronomy, its worth taking a look at the SATCON-1 webinar [URL136]. Shorter but more technical is this talk by Tony Tyson [URL137] from the Rubin Observatory project. For an explanation of what SpaceX are up to, and why they want thousands of spacecraft in Low Earth Orbit, this video by Real Engineering [URL138] is very good. For more general background, here are two good documentaries: this one [URL139] is a broad view of the "New Space Race", comparing the differing styles and ambitions of Elon Musk and Jeff Bezos; and this one [URL140] concentrates more specifically on SpaceX. One thing I understood better from these documentaries is that although Elon Musk was a rich man by normal standards when he started SpaceX, he had nowhere near the almost bottomless pockets that Jeff Bezos does. It has therefore been especially important for SpaceX to make money and win contracts as it goes along, and Starlink is part of that business plan; it is needed as a cash generator, so that SpaceX can go to Mars.

I only recently stumbled across a series of videos by Mark Hadley about Starlink from the point of view of network scientists. This video [URL141] speculates that although the Starlink constellation was originally designed to use laser communication between satellites, this has not yet been achieved; instead the initial incarnation will bounce signals up and down between satellites and a network of ground stations, including some ship-based ones. This won't deliver true global connectivity, but for connections between major financial centres, will still give the super-low latency that high-frequency trading requires.

Finally, take a look this video [URL142] of the drones launched over Scotland as part of Edinburgh's 2020 Hogmanay. Its enchanting; but now imagine the pretty lights spelling out the name of a popular chocolate bar. Its the beauty and the horror, right there in one video.

Petitions

LAST BUT NOT LEAST, two internet petitions have emerged on the internet. I strongly urge you sign one of them. If you are a professional astronomer, the appropriate one to sign is this one [URL81], called "Safeguarding the Astronomical

Sky", and set up by Stefano Gallozi. It contains some quite specific suggestions about pursuing legal avenues. It has been signed by over two thousand astronomers worldwide, but we could do with more.

If on the other hand you are a member of the general public, the best one to sign is this one [URL143], called "Losing the night sky due to tens of thousands of satellites", and started by "Gaspar B." on the Avaaz.org community petition site. Last time I looked, 16,787 people had signed this petition, but I am sure we could get it even higher. The stated aim is "to help us tell the Federal Communications Commission not to give permits (or withdraw such permits) to SpaceX and other companies for launching additional satellites that will potentially harm our view of the night sky". This petition also says "**It is only a matter of a few years before the sky as we know it - and as our ancestors have known it for millenia - will be irretrievably lost**".

Go sign it.

BIBLIOGRAPHY

Here are the sources referred to in the Appendix. I have divided them into two parts - firstly a list of URLs, i.e. links to web pages and youtube videos; and secondly a list of books and articles - all of which are also available online.

Web Links

[URL1] SpaceX website:
https://www.spacex.com

[URL2] Starlink website:
https://www.starlink.com

[URL3] OneWeb website:
https://oneweb.world

[URL4] OneWeb satellites:
https://onewebsatellites.com

[URL5] Telesat website:
https://www.telesat.com

[URL6] China Aerospace Corporation:
http://www.casic.com

[URL7] Roscosmos website:
http://en.roscosmos.ru

[URL8] Globalstar website:
https://www.globalstar.com/en-gb/

[URL9] Article by Larry Press in CircleID about Chinese plans:
http://www.circleid.com/posts/20201002-a-new-chinese-broadband-satellite-constellation

[URL10] Article in TASS about Russian plans:
https://tass.com/science/1213791

[URL11] SpaceNews article about Amazon plans:
https://spacenews.com/amazon-planning-3236-satellite-constellation-for-internet-connectivity/

[URL12] Wikipedia entry about Amazon's Kuiper project:
https://en.wikipedia.org/wiki/Kuiper_Systems

[URL13] Lynk website:
https://lynk.world

[URL14] Guardian report on failed Facebook launch:
https://www.theguardian.com/commentisfree/2016/sep/05/facebook-satellite-developing-world-mark-zuckerberg-internet

[URL15] Article in "Advanced Television" about the Facebook experiment:
https://advanced-television.com/2020/10/14/facebook-enters-satellite-business/

[URL16] AGI Youtube video about the 107,00 proposed low orbit satellites:
https://www.youtube.com/watch?reload=9&v=oWB7ZySDHg8&feature=youtu.be

[URL17] AGI website:
https://www.agi.com

[URL18] ITU webinar on low-orbit constellations:
https://www.itu.int/en/ITU-R/space/workshops/sat-webinars/Pages/default.aspx

[URL19] Viasat internet website:
https://www.viasat.com/internet

[URL20] O3B website:
https://www.ses.com/networks/networks-and-platforms/o3b-meo

[URL21] SES parent website:
https://www.ses.com

[URL22] Wikipedia entry on O3B system:
https://en.wikipedia.org/wiki/O3b_MEO

[URL23] OneWeb Responsible Space website:
https://www.oneweb.world/responsible-space

[URL24] Gnosis Network on sustainable space:
https://gnosisnetwork.org

[URL25] Satellite Industry Association:
https://sia.org

[URL26] Space Data Association:
https://www.space-data.org/sda/

[URL27] Satcom Innovations Group:
https://satig.space

[URL28] EMEA Satellite Operators Association:
https://www.esoa.net

[URL29] Global VSAT Forum:
https://gvf.org

[URL30] Celestrak website:
http://celestrak.com

[URL31] Wikipedia entry on 18th Space Control Squadron:
https://en.wikipedia.org/wiki/18th_Space_Control_Squadron

[URL32] Wikipedia entry on US Space Surveillance Network:
https://en.wikipedia.org/wiki/United_States_Space_Surveillance_Network

[URL33] Space Track website:
https://www.space-track.org

[URL34] Russian Space Surveillance programme:
http://spacedata.vimpel.ru

[URL35] Leolabs website:
https://www.leolabs.space

[URL36] Exo-analytics website:
https://exoanalytic.com

[URL37] Wikipedia entry on amateur satellite tracking:
https://en.wikipedia.org/wiki/Satellite_watching

[URL38] Planet4589 website, run by Jonathan McDowell:
https://planet4589.org

[URL39] Jonathan's Space Report:
https://planet4589.org/space/jsr/jsr.html

[URL40] McDowell's General Catalog of Space Objects:
https://planet4589.org/space/gcat/index.html

[URL41] Satview website:
https://www.satview.org

[URL42] N2YO satellite tracking website:
https://www.n2yo.com

[URL43] In The Sky website:
https://in-the-sky.org

[URL44] European Space Agency (ESA) website:
https://www.esa.int

[URL45] US National Aeronautics and Space Administration (NASA) website:
https://www.nasa.gov

[URL46] Japanese Aerospace Exploration Agency (JAXA):
https://global.jaxa.jp

[URL47] Russian Space Agency (Roscosmos):
http://en.roscosmos.ru

[URL48] Space News website:
https://spacenews.com

[URL49] Spaceflight Now website:
https://spaceflightnow.com

[URL50] Space.com website:
https://www.space.com

[URL51] Spacewatch global website:
https://spacewatch.global

[URL52] Skyweek website:
https://skyweek.wordpress.com

[URL53] Space Policy Online:
https://spacepolicyonline.com

[URL54] Business Insider article arbout SpaceX launch failures:
https://www.businessinsider.com/spacex-starlink-internet-satellites-percent-failure-rate-space-debris-risk-2020-10?r=US&IR=T

[URL55] Planet4589 Twitter Account:
https://twitter.com/planet4589

[URL56] Mark McCaughrean Twitter Account:
https://twitter.com/markmccaughrean

[URL57] Wikipedia entry on Space Debris:
https://en.wikipedia.org/wiki/Space_debris

[URL58] ESA Space Debris home page:
http://www.esa.int/Safety_Security/Space_Debris

[URL59] ESA Space Debris Video playlist:
http://www.esa.int/ESA_Multimedia/Sets/Space_debris_playlist/(result_type)/videos

[URL60] ESA page on Current State of Space Debris:
http://www.esa.int/Safety_Security/Space_Debris/The_current_state_of_space_debris

[URL61] Space Debris pages of European Centre for Space Law:
https://www.esa.int/About_Us/ECSL_-_European_Centre_for_Space_Law/Space_Debris3

[URL62] ESA Space Debris User Portal:
https://sdup.esoc.esa.int

[URL63] Arstechnica article about planned Russian Sphere system:
https://arstechnica.com/science/2021/01/russia-may-fine-citizens-who-use-spacexs-starlink-internet-service/

[URL64] NASA Orbital Debris Program Office (ODPO):
https://orbitaldebris.jsc.nasa.gov

[URL65] NASA Offfice of Safety and Mission Assurance (OSMA) Orbital Debris pages:
https://sma.nasa.gov/sma-disciplines/orbital-debris

[URL66] ODPO Quarterly Debris newsletter:
https://orbitaldebris.jsc.nasa.gov/quarterly-news/#

[URL67] ODPO standard practices reference document:
https://orbitaldebris.jsc.nasa.gov/library/usg_orbital_debris_mitigation_standard_practices_november_2019.pdf

[URL68] ODPO modelling page, including a video on growth of Debris population over time:
https://orbitaldebris.jsc.nasa.gov/modeling/

[URL69] ODPO protection page, including a video with a 3D fly-through of the modelled population:
https://orbitaldebris.jsc.nasa.gov/protection/

[URL70] Wikipedia entry on the NASA Orbital Debris Program Office (ODPO):
https://en.wikipedia.org/wiki/NASA_Orbital_Debris_Program_Office

[URL71] Don Kessler's personal web page:
http://aquarid.physics.uwo.ca/kessler/iwp5.html

[URL72] Compendium of national debris mitigation deadlines compiled by COPUOS:
https://www.unoosa.org/oosa/en/ourwork/topics/space-debris/compendium.html

[URL73] Wesbite for IADC 2020 meeting:
https://iadc2020.com

[URL74] COPUOS Debris Mitigation Guidelines document:
https://www.unoosa.org/res/oosadoc/data/documents/2010/stspace/stspace49_0_html/st_space_49E.pdf

[URL75] "The Elusive Peril of Space Junk" by Raffi Khatchadourian, New Yorker September 2020: https://www.newyorker.com/magazine/2020/09/28/the-elusive-peril-of-space-junk

[URL76] Wikipedia entry on the Kessler Syndrome: https://en.wikipedia.org/wiki/Kessler_syndrome

[URL77] EathData article by Agnieszka Gautier on Debris tracking: https://earthdata.nasa.gov/learn/sensing-our-planet/rubble-trouble

[URL78] CNN online article by Jackie Wattles on a recent near collision: https://edition.cnn.com/2020/10/16/tech/space-junk-tracking-scn/index.html

[URL79] Space News article by Jeff Foust on attitude of mega-constellations companies to sustainability of the space environment: https://spacenews.com/can-satellite-megaconstellations-be-responsible-users-of-space/

[URL80] Business Insider article by Dave Mosher about near collision between the ESA Aeolus satellite and Starlink 44: https://www.businessinsider.com/spacex-didnt-move-starlink-satellite-after-crash-warning-due-to-bug-2019-9?r=US&IR=T

[URL81] "Safeguarding the Sky", online petition: https://astronomersappeal.wordpress.com

[URL82] Satellite Constellations Working Group of the International Astronomical Union: https://iau.org/public/themes/satellite-constellations/

[URL83] Live updates page of the American Astronomical Society on the Impacts of Large Satellite Constellations on Astronomy:
https://aas.org/posts/advocacy/2020/12/impacts-large-satellite-constellations-astronomy-live-updates

[URL84] Royal Astronomical Society (UK) working group on satellite megaconstellations:
https://ras.ac.uk/ras-policy/science-policy-consultative-forums/satellite-megaconstellations

[URL85] Web page for Dark and Quiet Skies conference, October 2020:
http://research.iac.es/congreso/quietdarksky2020/

[URL86] Website of the International Dark-Sky Association (IDA):
https://www.darksky.org

[URL87] Galloway Forest Dark Sky Park:
https://forestryandland.gov.scot/visit/forest-parks/galloway-forest-park/dark-skies

[URL88] Dark Sky Scotland website:
https://www.darkskyscotland.org.uk/dark_sky_places/

[URL89] Globe at Night citizen science project:
https://www.globeatnight.org

[URL90] "Protecting our planet at night", online presentation by Diana Umpierre:
https://app.box.com/s/0h02lshu04cj9rq7z63d5fp4pg2dwcb5

[URL91] Journal of Dark Sky Studies:
https://epubs.utah.edu/index.php/jdss/index

[URL92] Twitter feed for John Barentine, Policy Director of IDA:
https://twitter.com/JohnBarentine

[URL93] All Party Parliamentary Group for Dark Skies (UK Parliament):
https://appgdarkskies.co.uk

[URL94] Wikipedia entry on International Law:
https://en.wikipedia.org/wiki/International_law

[URL95] Wikipedia entry on class action lawsuits:
https://en.wikipedia.org/wiki/Class_action

[URL96] Article about the new EU collective legal action laws, o the Deutsche Welle website:
https://www.dw.com/en/class-action-lawsuits-to-become-eu-law/a-55711222

[URL97] UNOOSA page about COPUOS, the Committee on the Peaceful Uses of Outer Space:
https://www.unoosa.org/oosa/en/ourwork/copuos/index.html

[URL98] Web page of UNOOSA, the United Nations Office of Outer Space Affairs:
https://www.unoosa.org

[URL99] UNOOSA Space Law FAQs:
https://www.unoosa.org/oosa/en/informationfor/faqs.html

[URL100] Handy summary of Space Law at Space Policy Online:
https://spacepolicyonline.com/topics/space-law/

[URL101] Web page of International Institute of Space Law:
https://iislweb.org

[URL102] Formal agreement between governments of Canada and USSR, at the UNOOSA website:
https://www.unoosa.org/oosa/en/ourwork/spacelaw/nationalspacelaw/bi-multi-lateral-agreements/can_ussr_001.html

[URL103] Report on "Operation Morning Light" at the Arctic Operational Histories website:
http://operationalhistories.ca/home/operation-morning-light/

[URL104] Wikipedia entry on the Antarctic Treaty System:
https://en.wikipedia.org/wiki/Antarctic_Treaty_System

[URL105] United Nations Division for Ocean Affairs and the Law of the Sea:
https://www.un.org/Depts/los

[URL106] Historical perspective on the development of the Law of the Sea, at the UN Division for Ocean Affairs website:
https://www.un.org/Depts/los/convention_agreements/convention_historical_perspective.htm#Historical%20Perspective

[URL107] Website of the International Seabed Authority:
https://www.isa.org.jm

[URL108] Website of the International Tribunal for the Law of the Sea:
https://www.itlos.org

[URL109] Website of the Internet Engineering Task Force (IETF):
https://www.ietf.org

[URL110] Website of the Internet Corporation for the Assignment of Names and Numbers (ICANN):
https://www.icann.org

[URL111] Website of the World Wide Web Consortium (W3C): https://www.w3.org

[URL112] UK government web page explaining how to apply for a space operations license: https://www.gov.uk/guidance/apply-for-a-license-under-the-outer-space-act-1986

[URL113] Home page of UK Space Agency: https://www.gov.uk/government/organisations/uk-space-agency

[URL114] Home page of International Telecommunication Union (ITU): https://www.itu.int

[URL115] Satellite filings page of OFCOM, the UK telecommunications regulator: https://www.ofcom.org.uk/spectrum/information/satellites-space-science/satellite-filings

[URL116] US Office of Commercial Space Transportation, part of the Federal Aviation Authority: https://www.faa.gov/licenses_certificates/commercial_space_transportation/

[URL117] Home page of US Federal Communications Commission: https://www.fcc.gov

[URL118] Web page for Radiocommunication sector of ITU: https://www.itu.int/en/ITU-R

[URL119] Video about the work of ITU-R: https://www.youtube.com/watch?v=ncNRTsNDrkg&feature=emb_rel_end

[URL120] Web page for Digital Inclusion sector of ITU:

https://www.itu.int/en/ITU-D/Digital-Inclusion

[URL121] Video about the work of ITU-D: https://www.youtube.com/watch?v=m0Kp139nMes&feature=youtu.be

[URL122] News item at CNN reporting $856M subsidy of Starlink by the US FCC: https://edition.cnn.com/2020/12/08/tech/spacex-starlink-subsidies-fcc-scn/index.html

[URL123] News item at Spacewatch reporting Viasat complaint to USS FCC about change of Starlink orbit plans: https://spacewatch.global/2020/12/viasat-concerned-about-spacexs-starlink-relocation-plans/

[URL124] Webinar at Spacewatch by Chris Stott about the permissions regime for space launches: https://spacewatch.global/2020/11/space-cafe-webtalk-recap-chris-stott-in-a-conversation-about-permissions-and-more/

[URL125] Webinar at Spacewatch by Ram Jakhu about the need for more space law: https://spacewatch.global/2020/10/space-cafe-webtalk-recap-prof-dr-ram-jakhu-on-the-need-for-more-space-laws/

[URL126] Article by Astronomer Royal Martin Rees in the Guardian, Dec 2019: https://www.theguardian.com/commentisfree/2019/dec/29/be-wary-of-elon-musk-despoiling-the-vault-of-heaven

[URL127] Video of double train of Starlink satellites ascending through sky, taken by Olivier Staiger: https://www.youtube.com/watch?v=FsySNQ8j2Tg

[URL128] Information about the DESC image shown in Fig. 5.2, from the website of the US National Optical and IR Lab (NOIRLab):
https://noirlab.edu/public/images/iotw1946a/

[URL129] The Hubble Space Telescope observation that Fig. 5.3 is taken from. This is a public preview image from the Mikulski Archive for Space Telescopes. The PI for the science data is Simon Porter:
https://archive.stsci.edu/cgi-bin/mastpreview?mission=hst&dataid=IEDK12AOQ

[URL130] The Guardian Astronomy Photographer of the Year (2020) competition, which features the image of Albireo taken by Rafael Schmall which is the cover image of this book:
https://www.theguardian.com/science/gallery/2020/sep/11/winners-royal-observatory-greenwich-astronomy-photographer-year-2020-in-pictures

[URL131] The Satellite Streak Watcher project:
https://www.anecdata.org/projects/view/687/photos

[URL132] Simulation video of naked eye view of 12,000 Starlink satellites, made by Michael Vlasov:
https://www.youtube.com/watch?v=z9hQfKd9kfA

[URL133] 360 VR simulation video of what the sky would look like if you could see all known satellites with your naked eye, made by Scott Manley:
https://www.youtube.com/watch?v=dJNGi-bt9NM

[URL134] Simulation video of what the Rubin Observatory would see, if it was operating today:
https://www.youtube.com/watch?v=nAaL85cBsp4

[URL135] AGI Spacecast 27, with T.S.Kelso explaining how satellites are tracked:
https://www.youtube.com/watch?v=1TSHc9GIFPI

[URL136] The SATCON-1 webinar, jointly hosted by the American Astronomical Society and the Satellite Industry Association:
https://www.youtube.com/watch?v=vAStpnfjipU&t=118s

[URL137] Talk given by Tony Tyson at the Rubin Observatory Project and Community Workshop 2020:
https://www.youtube.com/watch?v=ulVRA3z7C1I&feature=youtu.be

[URL138] "Why SpaceX is making Starlink", and explanatory video in the "Real Engineering" series:
https://www.youtube.com/watch?v=giQ8xEWjnBs

[URL139] "The New Space Race of the 2020's", a documentary by ColdFusion:
https://www.youtube.com/watch?v=bqYbgWbKMPc

[URL140] "SpaceX: Capitalism at its Finest", a documentary made by Jake Tran:
https://www.youtube.com/watch?v=0JxchCdt4gg

[URL141] Video by Mark Handley speculating on how the Starlink ground relay system may work:
https://www.youtube.com/watch?v=m05abdGSOxY&pbjreload=101

[URL142] Webpage of "Edinburgh's Hogmanay 2020", featuring a video showing the drones launched over Edinburgh and other parts of Scotland:
https://www.edinburghshogmanay.com

[URL143] "Losing the night sky due to tens of thousands of satellites", a petition started by "Gaspar B": https://secure.avaaz.org/community_petitions/en/SpaceX_and_companies_planning_to_launch_constellations_of_satellites_Losing_the_night_sky_due_to_tens_of_thousands_of_satellites/?s=03

[URL144] The movie referred to in Chapter 1: "A Residence Above the Skies". This link has a short trailer, but also explains the availability of the movie through Chip Taylor Communications: https://www.chiptaylor.com/residence_above_clouds.html

References to key documents, reports, research papers, and books

[R1] "Annual Space Environment Report". Prepared by: ESA Space Debris Office. Reference: GEN-DB-LOG-00288-OPS-SD. Issue/Version: 4.0. Issue date: 29 September 2020. *Accessible at:* https://www.sdo.esoc.esa.int/environment_report/Space_Environment_Report_latest.pdf

[R2] "Study on space environment safety based on satellite collision" Honquiang,S., and Zhanyue, Z., 2020. IOP Conference Series: Earth and Environmental Science, Volume 552, p.012014. *Accessible at:* https://ui.adsabs.harvard.edu/abs/2020E%26ES..552a2014H/abstract

[R3] "DebrisWatch I: A survey of faint geosynchronous debris", Blake, J.A. et al, 2021. Advances in Space Research, Volume 67, p.360. *Accessible at:* https://ui.adsabs.harvard.edu/abs/2021AdSpR..67..360B/abstract

[R4] "The Low Earth Orbit Satellite Population and Impacts of the SpaceX Starlink Constellation", McDowell, J., 2020. Astrophysical Journal, Volume 892, L36. *Accessible at:* https://ui.adsabs.harvard.edu/abs/2020ApJ...892L..36M/abstract

[R5] "Concerns about ground based astronomical observations: a step to safeguard the astronomical sky", Gallozzi, A., Scardia, M., and Maris, M., 2020. IOP Conference Series: Earth and Environmental Science, Volume 552, p.012014. *Accessible at:* https://ui.adsabs.harvard.edu/abs/2020arXiv200110952G/abstract

[R6] "Concerns about ground-based astronomical observations: quantifying satellites constellations damages", Gallozzi, S. et al, 2020. ArxiV 2003.05472. *Accessible at:* https://ui.adsabs.harvard.edu/abs/2020arXiv200305472G/abstract

[R7] "Dark skies and bright satellites", Tyson, A., and Parriott, J., 2020. Science, Vol. 369, p.1543. *Accessible at:* https://science.sciencemag.org/content/369/6511/1543/tab-article-info

[R8] "Impact of satellite constellations on astronomical observations with ESO telescopes in the visible and infrared domains", Hainaut, O.R., and Williams, A.P., 2020. Astronomy and Astrophysics, Volum 636, A121. *Accessible at:* https://ui.adsabs.harvard.edu/abs/2020A%26A...636A.121H/abstract

[R9] "Mitigation of LEO Satellite Brightness and Trail Effects on the Rubin Observatory LSST", Tyson, J.A., et al 2020. Astronomical Journal, Volume 160, p.226 *Accessible at:* https://ui.adsabs.harvard.edu/abs/2020AJ....160..226T/abstract

[R10] "First observations and magnitude measurement of Starlink's Darksat", Tregloan-Reed, J. et al, 2020a. Astronomy and Astrophysics, Volume 637, L1. *Accessible at:* https://ui.adsabs.harvard.edu/abs/2020A%26A...637L...1T/abstract

[R11] "Optical to NIR magnitude measurements of the Starlink LEO Darksat satellite and effectiveness of the darkening treatment", Tregloan-Reed, J., et al 2020b. Astronomy and Astrophysics under review. *Accessible at*: https://ui.adsabs.harvard.edu/abs/2020arXiv201101820T/abstract

[R12] "The Brightness of OneWeb Satellites", Mallama, A., 2020. Preprint, ArXiV 2012.05100. *Accessible at*: https://ui.adsabs.harvard.edu/abs/2020arXiv201205100M/abstract

[R13] "Satellite Constellation Internet Affordability and Need", Rawls, M.L., et al 2020. Research Notes of the American Astronomical Society, Volume 4, p.189. *Accessible at*: https://ui.adsabs.harvard.edu/abs/2020RNAAS...4..189R/abstract

[R14] "Impact of Satellite Constellations on Optical Astronomy and Recommendations Toward Mitigations", Walker, C. et al, 2020. Bulletin of the American Astronomical Society, Vol. 52, No. 2. *Accessible at*:
https://ui.adsabs.harvard.edu/abs/2020BAAS...52.0206W/abstract

[R15] "On-line Workshop: Dark and Quiet Skies for Science and Society: Report and recommendations", Walker, C., et al, 2020. NOIRLab Technical Document, ID techdoc021, release date Jan 11th 2021. *Accessible here*:
https://noirlab.edu/public/products/techdocs/techdoc021/

[R16] "Recommendations to Keep Dark and Quiet Skies for Science and Society", 2020. Submission to Scientific and Technical Subcommittee Fifty- eighth session, April 2021, COPUOS. *Accessible here*:
https://www.iau.org/static/publications/uncopuos-stsc-crp-8jan2021.pdf

[R17] "Space Activities in 2020", Jonathan McDowell's latest annual report *Accessible here*:
https://planet4589.org/space/papers/space20.pdf

[R18] "Effects of the COVID-19 lockdown on urban light emissions: ground and satellite comparison", Bustamante-Calabria, M., et al 2021. Remote Sensing, Volume 13, p.258. *Accessible here*:
https://arxiv.org/abs/2011.09252

[R19] "The economics of global light pollution", Gallaway, T., Olsen, R.N., and Mitchell, D.M., 2010. Ecological Economics, Volume 69, p.658. *Accessible here*:
https://www.sciencedirect.com/science/article/pii/S0921800909004121

[R20] "Direct measurement of the contribution of street lighting to satellite observations of nighttime light emissions from urban areas", Kyba, C.C.M., *et al*, 2020. Lighting Research and Technology, Volume 0, p.1. *Accessible here*: https://journals.sagepub.com/doi/10.1177/1477153520958463

[R21] "Ten Dark Sky Policies for the Government", report submitted to UK Government. Authored by Chris Cook on behalf of members of All Party Parliamentary Group for Dark Skies, 2020. *Accessible here*: https://static1.squarespace.com/static/5e567fb65a380a76eb3c8133/t/5ffd7eee3584326aea8a794d/1610448653257/Ten+Dark+Sky+policies+for+the+government+-+APPG+for+Dark+Skies.pdf

[R22] "International Space Law: United Nations Instruments", 2017. Published by United Nations Office of Outer Space Affairs. *Accessible here*: https://www.unoosa.org/res/oosadoc/data/documents/2017/stspace/stspace61rev_2_0_html/V1605998-ENGLISH.pdf

[R23] "The Little Book of Space Law", Kleiman, M.J., 2014. Published by the American Bar Association. *Kindle and paperback versions available at Amazon*:

[R24] "Making policy for new asteroid activities: In pursuit of science, settlement, security, or sales?", Krolikowski, A., and Elvis, M., 2019. Space Policy, Volume 47, p.7. *Available here*: https://www.sciencedirect.com/science/article/abs/pii/S0265964618300262

[R25] "Concentrated lunar resources: imminent implications for governance and justice", Elvis. M., Krolikowski, A., and Milligan, T., 2020. Phil.Trans.R.Soc.A., Volume 378, p. 20190563. *Available here*: https://royalsocietypublishing.org/doi/10.1098/rsta.2019.0563

[R26] "Asteroids: How Love, Fear, and Greed Will Determine "Our Future in Space", Elvis., M, 2021. Published by Yale University Press. Hardback available at Amazon"

Acknowledgements

The two most important people to thank are Ski and Jonathan. It was Ski Antonucci who first had the idea of such a book, convinced me to do it once I seemed interested, and then kept me encouraged and fed with interesting links over multiple chapter drafts. Meanwhile Jonathan McDowell agreed to be my technical consultant, and responded very quickly to all my questions. It would be a much weaker book without Jonathan. It also got early encouragement that the idea was a good thing, from a local Edinburgh multi-disciplinary group called "Social Dimensions of Science", especially Pippa Goldschmidt, Jarita Holbrook, Mike Nellis, Deborah Scott, and Matjaz Vidmar. I also had early helpfull comments and suggestions from colleagues at the Astronomical Society of Edinburgh - especially Andrew Farrow, Graham Rule, and Sean Wixted. Among my professional colleagues, as well as Ski and Jonathan, I had useful suggestions from Mike Watson, Martin Elvis, Mark McCaughrean, and Meredith Rawls. I am also grateful to all the people who have allowed me to use their images - all named checked in the figure captions. I am especially grateful to Rafael Schmall, whose prize-winning picture I have used for the cover, and Gordon Mackie, who took what I think is my favourite ever Milky Way picture. Many thanks also to Brian May, who, as well as agreeing to write a foreword, made some very useful critical suggestions on the text, and also to Brian's publisher at the London Stereoscopic Company, Robin Rees. Last but very much not least I want to thank my partner Emma Griffiths, who read various stages of the book, and made numerous insightful suggestions while I was spilling my thoughts at the dinner table.

This book was put together and assembled into multiple formats using the rather wonderful Jutoh software package: http://www.jutoh.com

About the Author

Andy Lawrence is the Regius Professor of Astronomy at the University of Edinburgh. He works at the very beautiful and historic Royal Observatory Edinburgh. His research is mostly on Quasars and Active Galactic Nuclei, but he has always been a leader in the world of survey astronomy. He has published over 300 scientific articles, along with two textbooks and a booklet about his famous Victorian predecessor, Charles Piazzi Smyth:

Probability in Physics: an Introductory Guide

Astronomical Measurement: a Concise Guide

Stars, Time and Mountains: the work of Charles Piazzi Smyth, 1819-1900

Andy's personal website can be found at : https://andyxlastro.me

Photon Productions is a name under which Andy trades.

Portrait by Max Alexander, as part of the "Explorers of the Universe" exhibition for the Science and Technology Facilities Council, 2009. You can find more marvelous pictures at http://www.maxalexander.com

Made in the USA
Monee, IL
08 November 2026

68156254R00090